A Practical Guide to Graphite Furnace Atomic Absorption Spectrometry

CHEMICAL ANALYSIS

A SERIES OF MONOGRAPHS ON
ANALYTICAL CHEMISTRY AND ITS APPLICATIONS

Editor
J. D. WINEFORDNER

VOLUME 149

A Wiley-Interscience Publication

JOHN WILEY & SONS, INC.

New York / Chichester / Weinheim / Brisbane / Singapore / Toronto

A Practical Guide to Graphite Furnace Atomic Absorption Spectrometry

DAVID J. BUTCHER

Western Carolina University
Cullowhee, NC

JOSEPH SNEDDON

McNeese State University
Lake Charles, LA

A Wiley-Interscience Publication

JOHN WILEY & SONS, INC.

New York / Chichester / Weinheim / Brisbane / Singapore / Toronto

Library of Congress Cataloging-in-Publication Data:

Butcher, David J.
A practical guide to graphite furnace atomic absorption spectrometry / David J. Butcher, Joseph Sneddon.
p. cm.—(Chemical analysis; v. 147)
"A Wiley-Interscience publication."
Includes bibliographical references and index.
ISBN 0-471-12553-9 (cloth : alk. paper)
1. Furnace atomic absorption spectroscopy. I. Sneddon, Joseph, 1951– . II. Title. III. Series.
QD96.A8B87 1998
543'.0858—dc21 97-29336

Printed in the United States of America.

10 9 8 7 6 5 4 3 2 1

CONTENTS

PREFACE

Graphite furnace atomic absorption spectrometry (GFAAS) is a widely used method of elemental analysis. During the development of the technique, GFAAS was regarded as a difficult technique to use, especially compared to flame atomic absorption. This sentiment was summarized by Herb Kahn (1), who said, "The difference between flame atomizers and carbon atomizers is like the difference between watching tennis and playing tennis." Since this statement was made, routine use of GFAAS has been greatly simplified by the use of a series of instrumental and methodological improvements that together are called modern furnace technology. The purpose of this volume is to provide an analyst with sufficient information, in conjunction with an atomic absorption cookbook, to use GFAAS for practical analysis. Topics include fundamental concepts of GFAAS, including theory, instrumentation, and methodology; procedures to develop practical methods of analysis; and types and costs of commercial GFAAS instrumentation. The focus is on practical applications of GFAAS; consequently, little emphasis is placed on laboratory-constructed instrumentation and other specialized concepts. This volume does not describe GFAAS applications in detail but instead provides the reader with a guide to the literature.

REFERENCE

1. J. W. Robinson, *Anal. Chem.* **66**, 472A (1994).

ACKNOWLEDGMENTS

We would like to thank Professor James D. Winefordner (University of Florida), who edits the Chemical Analysis Series, for the invitation to write this book. D.J.B. would like to acknowledge the Department of Chemistry and Physics at Western Carolina University, which provided release time to work on this project. D.J.B. would also like to acknowledge the assistance of his wife, Karen Butcher, who proofread the manuscript and offered numerous suggestions. J.S. would like to acknowledge the generous support of Thermo Jarrell Ash Corporation, in particular, Gerald R. Dulude, Zach Moseley, and John J. Sotera, and the Louisiana Education Quality Support Fund (LEQSF) Research program for 1994–1996 (Grant RD-A-21). We would also like to thank the staff at John Wiley and Sons for their assistance, particularly Editor Carla A. Fjerstad. Finally, we would like to acknowledge the anonymous efforts of our reviewers, who contributed significantly to the content and style of the volume.

DAVID J. BUTCHER
Western Carolina University

JOSEPH SNEDDON
McNeese State University

CHEMICAL ANALYSIS

A SERIES OF MONOGRAPHS ON ANALYTICAL CHEMISTRY AND ITS APPLICATIONS

J. D. Winefordner, *Series Editor*

Vol. 1. **The Analytical Chemistry of Industrial Poisons, Hazards, and Solvents**. *Second Edition*. By the late Morris B. Jacobs

Vol. 2. **Chromatographic Adsorption Analysis**. By Harold H. Strain (*out of print*)

Vol. 3. **Photometric Determination of Traces of Metals**. *Fourth Edition*
Part I: General Aspects. By E. B. Sandell and Hiroshi Onishi
Part IIA: Individual Metals, Aluminum to Lithium. By Hiroshi Onishi
Part IIB: Individual Metals, Magnesium to Zirconium. By Hiroshi Onishi

Vol. 4. **Organic Reagents Used in Gravimetric and Volumetric Analysis**. By John F. Flagg (*out of print*)

Vol. 5. **Aquametry: A Treatise on Methods for the Determination of Water**. *Second Edition (in three parts)*. By John Mitchell, Jr. and Donald Milton Smith

Vol. 6. **Analysis of Insecticides and Acaricides**. By Francis A. Gunther and Roger C. Blinn (*out of print*)

Vol. 7. **Chemical Analysis of Industrial Solvents**. By the late Morris B. Jacobs and Leopold Schetlan

Vol. 8. **Colorimetric Determination of Nonmetals**. *Second Edition*. Edited by the late David F. Boltz and James A. Howell

Vol. 9. **Analytical Chemistry of Titanium Metals and Compounds**. By Maurice Codell

Vol. 10. **The Chemical Analysis of Air Pollutants**. By the late Morris B. Jacobs

Vol. 11. **X-Ray Spectrochemical Analysis**. *Second Edition*. By L. S. Birks

Vol. 12. **Systematic Analysis of Surface-Active Agents**. *Second Edition*. By Milton J. Rosen and Henry A. Goldsmith

Vol. 13. **Alternating Current Polarography and Tensammetry**. By B. Breyer and H. H. Bauer

Vol. 14. **Flame Photometry**. By R. Herrmann and J. Alkemade

Vol. 15. **The Titration of Organic Compounds** (*in two parts*). By M. R. F. Ashworth

Vol. 16. **Complexation in Analytical Chemistry: A Guide for the Critical Selection of Analytical Methods Based on Complexation Reactions**. By the late Anders Ringbom

Vol. 17. **Electron Probe Microanalysis**. *Second Edition*. By L. S. Birks

Vol. 81. **The Analysis of Extraterrestrial Materials**. By Isidore Adler

Vol. 82. **Chemometrics.** By Muhammad A. Sharaf. Deborah L. Illman, and Bruce Kowalski

Vol. 83. **Fourier Transform Infrared Spectrometry**. By Peter R. Griffiths and James A. de Haseth

Vol. 84. **Trace Analysis: Spectroscopic Methods for Molecules**. Edited by Gary Christian and James B. Callis

Vol. 85. **Ultratrace Analysis of Pharmaceuticals and Other Compounds of Interest**. Edited by S. Ahuja

Vol. 86. **Secondary Ion Mass Spectrometry: Basic Concepts, Instrumental Aspects, Applications and Trends.** By A. Benninghoven, F. G. Rüdenauer, and H. W. Werner

Vol. 87. **Analytical Applications of Lasers**. Edited by Edward H. Piepmeier

Vol. 88. **Applied Geochemical Analysis** By C. O. Ingamells and F. F. Pitard

Vol. 89. **Detectors for Liquid Chromatography**. Edited by Edward S. Yeung

Vol. 90. **Inductively Coupled Plasma Emission Spectroscopy: Part I: Methodology, Instrumentation, and Performance; Part II: Applications and Fundamentals**. Edited by J. M. Boumans

Vol. 91. **Applications of New Mass Spectrometry Techniques in Pesticide Chemistry**. Edited by Joseph Rosen

Vol. 92. **X-Ray Absorption: Principles, Applications, Techniques of EXAFS, SEXAFS, and XANES**. Edited by D. C. Konnigsberger

Vol. 93. **Quantitative Structure-Chromatographic Retention Relationships**. By Roman Kaliszan

Vol. 94. **Laser Remote Chemical Analysis**. Edited by Raymond M. Measures

Vol. 95. **Inorganic Mass Spectrometry**. Edited by F. Adams, R. Gijbels, and R. Van Grieken

Vol. 96. **Kinetic Aspects of Analytical Chemistry**. By Horacio A. Mottola

Vol. 97. **Two-Dimensional NMR Spectroscopy**. By Jan Schraml and Jon M. Bellama

Vol. 98. **High Performance Liquid Chromatography**. Edited by Phyllis R. Brown and Richard A. Hartwick

Vol. 99. **X-Ray Fluorescence Spectrometry**. By Ron Jenkins

Vol. 100. **Analytical Aspects of Drug Testing**. Edited by Dale G. Deutsch

Vol. 101. **Chemical Analysis of Polycyclic Aromatic Compounds**. Edited by Tuan Vo-Dinh

Vol. 102. **Quadrupole Storage Mass Spectrometry**. By Raymond E. March and Richard J. Hughes

Vol. 103. **Determination of Molecular Weight**. Edited by Anthony R. Cooper

Vol. 104. **Selectivity and Detectability Optimizations in HPLC**. By Satinder Ahuja

Vol. 105. **Laser Microanalysis**. By Lieselotte Moenke-Blankenburg

Vol. 106. **Clinical Chemistry**. Edited by E. Howard Taylor

Vol. 107. **Multielement Detection Systems for Spectrochemical Analysis**. By Kenneth W. Busch and Marianna A. Busch

Vol. 108. **Planer Chromatography in the Life Science**. Edited by Joseph C. Touchstone

Vol. 109. **Fluorometric Analysis in Biomedical Chemistry: Trends and Techniques Including HPLC Applications**. By Norio Ichinose, George Schwedt, Frank Michael Schnepel, and Kyoko Adochi

Vol. 110. **An Introduction to Laboratory Automation**. By Victor Cerdá and Guillermo Ramis

Vol. 141. **Spot Test Analysis: Clinical, Enviromental, Forensic, and Geochemical Applications**, Second Edition. By Ervin Jungreis

Vol. 142. **The Impact of Stereochemistry on Drug Development and Use**. Edited by Hassan Y. Aboul-Enein and Irving W. Wainer

Vol. 143. **Macrocyclic Compounds in Analytical Chemistry**. Edited by Yury A. Zolotov

Vol. 144. **Surface-Launched Acoustic Wave Sensors: Chemical Sensing and Thin-Film Characterization**. By Micheal Thompson and David Stone

Vol. 145. **Modern Isotope Ratio Mass Spectrometry**. By I. T. Platzner

Vol. 146. **High Performance Capillary Electrophoresis: Theory, Techniques, and Applications**. Edited by Morteza G. Khaledi

Vol. 147. **Solid-Phase Extraction: Principles and Practice**. By E.M. Thurman and M.S. Mills

Vol. 148. **Commercial Biosensors: Applications to Clinical, Bioprocess, and Environmental Samples**. By Graham Ramsay

Vol. 149. **A Practical Guide to Graphite Furnace Atomic Absorption Spectrometry**. By David J. Butcher and Joseph Sneddon

A Practical Guide to Graphite Furnace Atomic Absorption Spectrometry

CHAPTER 1

INTRODUCTION

Atomic absorption involves a measurement of the reduction of intensity of optical electromagnetic radiation, from a light source, following its passage through a cell containing gaseous atoms (the atom cell). Atomic absorption spectroscopy generally refers to the study of fundamental principles of this phenomenon, whereas atomic absorption spectrometry (AAS) refers to its use for the quantitative determination of elements in samples, although these terms are often used interchangeably. Atomic absorption spectrometry is applicable for the determination of most elements (almost all metals and metalloids and some nonmetals) in a wide variety of samples, including biological, clinical, environmental, food, and geological, and hence is one of the most commonly used techniques for elemental analysis. Two types of atom cells have been commonly used for AAS. The flame is widely used because of its ease of use for elemental analysis at the part per million (μg/mL) level. However, the use of a graphite furnace as the atomizer is used when a limited sample volume is available or lower analyte concentrations (part per billion, ng/mL level) are present. In this case, the technique is commonly referred to as graphite furnace atomic absorption spectrometry (GFAAS).

As the title indicates, the purpose of this book is to provide the reader with a practical knowledge of the basic principles, nomenclature, methodology, and applications of GFAAS. Basic principles of GFAAS are discussed in chapters on theory, instrumentation, interferences, and sample preparation/introduction. Concluding chapters discuss practical hints on the development of methods of analysis by GFAAS, including quality control procedures, characteristics of commercial instrumentation, and the future of the technique. Appendices include the historical background of the technique, some of its most relevant literature, and a compilation of conditions for analysis.

The success of GFAAS for elemental analysis involves the use of a series of instrumental developments and analytical protocols that together are called *modern furnace technology*. The components of modern furnace technology, along with their functions and the section where discussed in this volume, are given in Table 1.1. One of our goals is to emphasize the use of this technology for GFAAS in order to obtain precise and accurate results.

Table 1.1 Instrumentation and Protocols Included in Modern Furnace Technology

Component	Improvement	Section Discussed
Integrated absorbance	More accurate characterization of transient signal than peak absorbance	2.8–2.9
Pyrolytically coated graphite tubes	Reduced analyte–graphite interactions	4.2.1
Transversely heated furnace	Increased isothermality compared to longitudinally heated furnaces	4.2.1
Autosampler	Improved precision compared to manual pipetting	4.2.2
Platform or probe atomization with fast heating rates during atomization	Atomization in a hot environment that reduces interferences	4.2.3
Fast electronics	More accurate characterization of transient signal	4.5
Use of modern methods of background correction (self-reversal and Zeeman)	More accurate correction for background	4.6
Chemical modifiers	Reduced interferences	5.1.2

CHAPTER 2

THEORY

2.1 SPECTROSCOPY

Spectroscopy (1,2) is defined as the interaction of electromagnetic radiation (light) with matter. Electromagnetic radiation is described as having both wave and particle properties. Wave properties include frequency (ν, hertz), wavelength (λ, meters), velocity, and amplitude. Light is also considered to be composed of particles called photons that have a characteristic energy (E, joules). The relationships between energy, frequency, and wavelength are given by:

$$E = h\nu = \frac{hc}{\lambda} \tag{2.1}$$

where c is the speed of light in a vacuum (2.99792×10^8 m/s) and h is Planck's constant (6.62608×10^{-34} J s).

The electromagnetic spectrum covers a wavelength range of over 14 orders of magnitude, including the gamma ray, X-ray, ultraviolet, visible, infrared, microwave, and radio frequency regions (Table 2.1). For atomic absorption spectrometry, we will focus on a relatively limited region of the spectrum between 180 and 900 nm (ultraviolet, visible, and near infrared). These wavelengths are involved in electronic transitions of valence electrons.

Table 2.1 Spectral Regions of the Electromagnetic Radiation

Spectral Region	Wavelength Range	Type of Transition
Gamma ray	< 0.005 nm	Nuclear
X-ray	0.005–10 nm	Inner electrons
Vacuum (far) ultraviolet	10–180 nm	Bonding and middle shell electrons
Near UV	180–400 nm	Valence electrons
Visible	400–700 nm	Valence electrons
Infrared	0.7–1000 μm	Molecular vibrations and rotations
Microwave	1–300 mm	Molecular rotations
Radio wave	> 300 mm	Nuclear spin

2.2 INTRODUCTION TO ATOMIC SPECTROSCOPY

Atomic spectroscopy (1–7) involves the interaction of light with gaseous atoms. A device that converts a sample into gaseous atoms is called an *atom cell*. Typical atom cells include flames, plasmas, and graphite furnaces.

There are three basic types of atomic spectroscopy: atomic emission, atomic absorption, and atomic fluorescence. In order to introduce these phenomena, we will initially consider an atom with only two electronic energy states, in which the ground (lowest energy) state is designated 0 and the excited state as 1. It can generally be assumed that under normal conditions the majority of atoms are in the ground state.

Atomic emission (AE) (Fig. 2.1) involves the transfer of energy, usually as heat, from the atom cell to the atom to promote a valence electron in the atom from the ground state to the excited state. The atom then may emit a photon and

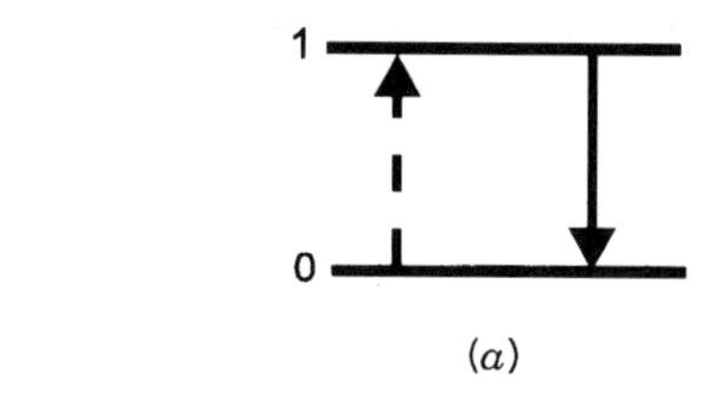

(*a*)

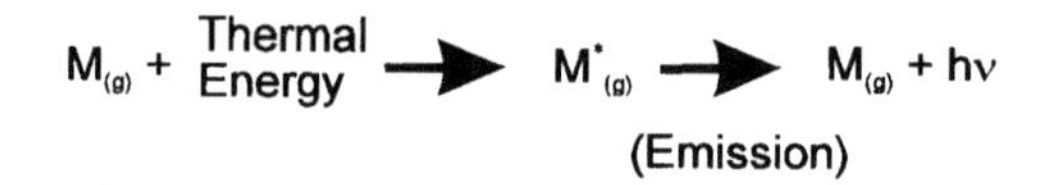

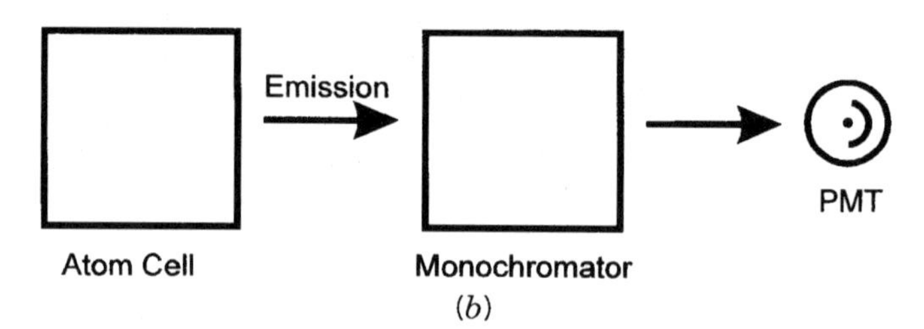

(*b*)

Figure 2.1. Atomic emission spectrometry: (*a*) transitions, where radiative transitions are represented by solid arrows (↓ = emission); nonradiative transitions by dashed arrows; $M_{(g)}$ represents ground-state analyte atoms and $M^*_{(g)}$ represents excited-state analyte atoms and (*b*) block diagram of instrumentation.

deactivate to the ground state (emission). The energy of the photon is equal to the difference in energy between the states. This process is called an *electronic transition*.

Atomic absorption (AA) (Fig. 2.2) involves the transfer of the energy of a photon to an atom (absorption) to promote a valence electron in the atom from the ground to the excited state. For absorption to occur, the energy of the photon must be identical to the difference in energy between the lower and higher energy levels of the atom.

Atomic fluorescence (AF) involves the excitation of atoms from a lower energy state (usually the ground state) to a higher energy state by light, followed by the emission (fluorescence) of a photon to deactivate the atoms (Fig. 2.3). Atomic fluorescence spectroscopy (AFS) can be considered a combination of atomic absorption and emission because it involves radiative excitation and deexcitation.

Atomic spectra are characterized by their relative simplicity, typically consisting of narrow lines, which correspond to the limited number of possible energy levels. Each element has a unique set of energy levels and hence a unique spectrum.

At the temperatures of the atom cells used for atomic absorption (1500–3000°C), the vast majority of atoms are present in the lowest energy level, called the ground state, and consequently the most sensitive lines involve transitions from the ground state, which are called *resonance transitions*. A transition from the lowest energy state is called a first resonance transition. The full width at half maximum (half-width) of most atomic lines is typically 0.01 to 0.05 nm, which is much narrower than liquid-phase molecular bands, whose widths are typically 10 to 100 nm. The absence of vibrational or rotational levels in atoms results in the narrow widths of atomic lines.

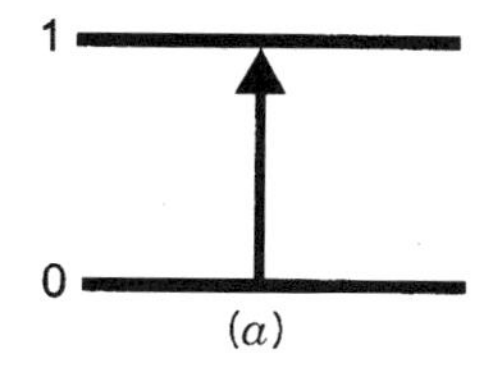

(*a*)

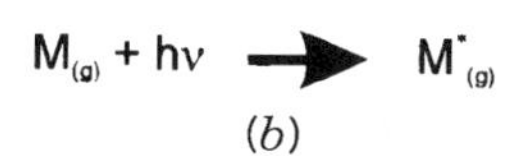

(*b*)

Figure 2.2. Atomic absorption transitions: (*a*) the promotion of an atom from a lower energy level (0) to a higher energy level (1) and (*b*) schematic representation of atomic absorption, where $M^{*}_{(g)}$ represents an excited-state atom.

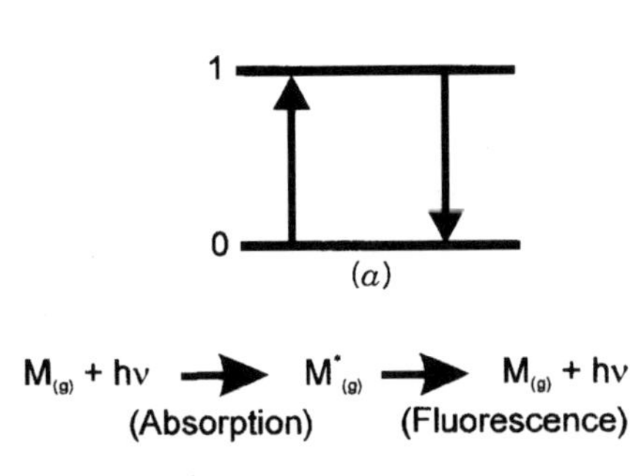

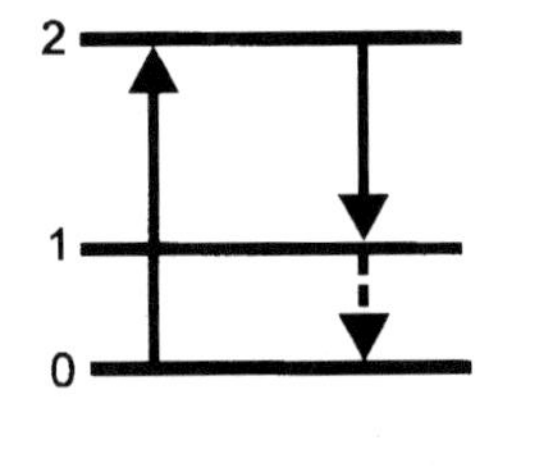

Figure 2.3. Transitions for AFS: (*a*) resonance and (*b*) nonresonance. The ground state is represented by 0; states 1 and 2 represent excited states. Radiative transitions are represented by solid arrows ($\uparrow$ = absorption; $\downarrow$ = fluorescence); nonradiative transitions by dashed arrows; $M_{(g)}$ represents ground-state analyte atoms and $M^{*}_{(g)}$ represents excited-state analyte atoms.

The remainder of this chapter deals with details of atomic spectroscopy that may be skipped by readers interested in practical analysis. For practical GFAAS analysis, a list of useful wavelengths for each element is provided in Appendix C and method books (atomic absorption cookbooks) provided with commercial instrumentation.

2.3 QUANTITATIVE CONSIDERATIONS OF TRANSITIONS

2.3.1 Population of Atoms in a State: The Boltzmann Distribution

For the simple two-level atoms illustrated in Figures 2.1 and 2.2, the ratio of atoms in excited state 1 to ground state 0 is given by the *Boltzmann distribution*:

$$\frac{N_1}{N_0} = \frac{g_1}{g_0} e^{-E/kT} \tag{2.2}$$

where N_1 and N_0 are the populations of the excited and ground states, respectively; g_1 and g_0 are the *statistical weights* of the excited and ground levels, which represent the number of degenerate states that make up that level (Section 2.4.2); E is the energy difference (J) between the states; k is Boltzmann's constant (1.3805×10^{-23} J/K); and T is the absolute temperature (K). For absorption spectroscopy, the maximum sensitivity is obtained when 100% of the atoms are in the ground state. Examination of the equation shows that the number of atoms in the ground state decreases with temperature and with low energy (long wavelength) transitions. The highest resonance wavelength for a commonly determined atom is 852.1 nm for cesium. Even for cesium at the maximum temperature employed in a graphite furnace (~3000 K), $N_1/N_0 = 0.007$. This result means that virtually all atoms formed in a graphite furnace are present in the ground state. This might allow one to assume that one could conclude that temperature has little effect on the number of atoms present, compared to emission spectroscopy, where changes in temperature may cause significant changes in N_1/N_0 (e.g., for Cs at 2000 K, $N_1/N_0 = 4 \times 10^{-4}$). This is true as long as there is no change in the fraction of atoms atomized. Temperature-dependent chemical reactions may occur that affect the number of gaseous atoms formed (chemical interferences).

2.3.2 Einstein Coefficients

In the simple model of an atom with two energy levels, three types of transitions are possible: absorption, in which a ground state atom is promoted to the excited state by interaction with a radiation field; *spontaneous emission*, in which an excited state atom spontaneously deactivates from the excited to the ground state; and *stimulated emission*, in which deactivation of the excited state is induced by the electromagnetic radiation. Einstein introduced three probability coefficients, B_{01}, A_{10}, and B_{10}, which are commonly called the *Einstein coefficients* for absorption, spontaneous emission, and stimulated emission, respectively, which can be used to evaluate radiative transition rates between the states (Fig. 2.4). The radiative rate of excitation per unit volume, $W_{0\to1}$ ($\text{cm}^{-3}\,\text{s}^{-1}$), depends only upon the rate of absorption and is given by:

$$W_{0\to1} = B_{01}\rho N_0 \tag{2.3}$$

where ρ is the energy density ($\text{J/cm}^3\,\text{Hz}$) of electromagnetic radiation at the frequency of the transition, N_0 is the population density (atoms/cm^3) of the ground state, giving the Einstein coefficient units of $\text{cm}^3\,\text{Hz/J s}$. The rate of radiative deactivation, $W_{1\to0}$ ($\text{cm}^{-3}\,\text{s}^{-1}$), is dependent on the rates of spontaneous emission and stimulated emission:

$$W_{1\to0} = A_{10}N_1 + B_{10}\rho N_1 \tag{2.4}$$

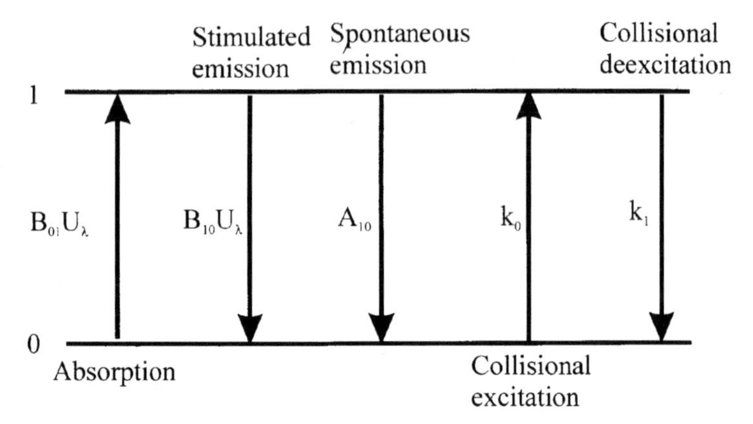

Figure 2.4. Activation and deactivation transitions in a two-level atom with corresponding rate constants. Absorption, stimulated emission, and spontaneous emission are radiative processes, and collisional excitation and deexcitation are nonradiative. The electromagnetic field has an energy density of U_λ at wavelength λ.

where N_1 is the population density of the excited state. The units of the Einstein coefficient for stimulated emission are identical to those for absorption, but those for spontaneous emission are s^{-1} since the rate is independent of the radiation field. The absorption and stimulated emission coefficients are related by:

$$B_{01}g_0 = B_{10}g_1 \tag{2.5}$$

while absorption and spontaneous emission coefficients follow this relationship:

$$A_{10} = \frac{8\pi h B_{01} g_0}{g_1 \lambda_m^2} \tag{2.6}$$

where g_1 and g_0 are the statistical weights of the excited and ground levels (Section 2.4.2); h = Planck's constant (6.626×10^{-34} J s); and λ_m is the wavelength of maximum absorption (emission). Einstein coefficients will be employed in Section 2.5 to define other measures of transition strength.

2.4 ELECTRONIC STATES OF ATOMS

The electronic states of atoms, and their resulting spectra, can be explained by quantum mechanics. Since atomic spectra involve valence electron transitions, the spectra depend on the possible energy states of these electrons. Quantum numbers are assigned to individual electrons, which in turn are used to assign a set of quantum numbers to the atoms.

2.4.1 Quantum Numbers for Electrons

According to quantum mechanics, electrons are assigned to locations called orbitals. The energy states of the orbitals are defined by four *quantum numbers*: n, the principal quantum number, which determines the energy of an orbital; l, the orbital angular momentum quantum number, which determines the shape of the orbital; m_l, the orbital magnetic quantum number, which describes the orientation of the orbital in space; and m_s, which defines the orientation of the spin angular momentum vector. The principal quantum number is a positive integer whose value specifies the value of l to be $0, 1, 2, \ldots, n-1$. The shapes of the orbitals are designated by the following scheme: $l = 0, s$; $l = 1, p$; $l = 2, d$; and $l = 3, f$. The energy of an electron is largely determined by n and l, and hence energy levels are commonly designated by the principal quantum number followed by the letter corresponding to the value of l, which is called a subshell (e.g., 1s, 2p, etc.). The orbital magnetic quantum number has integral values between $-l$ and $+l$. Each value of m_l represents a unique orbital. The spin quantum number may have values of $\frac{1}{2}$ or $-\frac{1}{2}$. In addition, the electron spin quantum number (s) represents the magnitude of the spin angular momentum, which has a value of $\frac{1}{2}$ for all electrons.

The *electron configuration* of an atom is specified by designation of principal and orbital angular momentum quantum numbers to each electron. The ground-state, or lowest energy configuration, can be determined from the maximum number of electrons per orbital and relative energies of the orbitals (lowest are filled first). The Pauli exclusion principle states that each electron in a given atom must have a unique set of quantum numbers, which implies that each orbital is limited to two electrons. Therefore an s shell may have up to two electrons; a p shell up to six; a d shell, 10; and an f shell, 14. The order of filling is described by the Aufbau order; $1s < 2s < 2p < 3s < 3p < 4s < 3d < 4p < 5s < 4d < 5p < 6s < 4f < 5d < 6p < 7s$. As an example, consider a sodium atom. A sodium atom has 11 electrons (equal to its atomic number Z), and electron configuration $1s^2 2s^2 2p^6 3s^1$.

Two additional quantum numbers are generated due to the interaction of orbital and spin angular momentum: j, the total angular momentum quantum number, with permitted values of $l + \frac{1}{2}$ and $l - \frac{1}{2}$ (except if $l = 0$, the only permitted value is $j = \frac{1}{2}$) and m_j, the total magnetic quantum number, with half-integer values between j and $-j$. Let us consider the example in which we have an s electron. For this case, $l = 0$ and $s = \frac{1}{2}$, so the only value is $j = \frac{1}{2}$. The total magnetic quantum number can have values of $\frac{1}{2}$ or $-\frac{1}{2}$.

2.4.2 Coupling Schemes, Term Symbols, and Selection Rules

In order to adequately describe a multielectron atom, it is necessary to define six additional quantum numbers (1,8): L, the resultant orbital angular momentum quantum number, with allowed values $l_1 + l_2, l_1 + l_2 - 1 \ldots |l_1 - l_2|$; M_L, the resultant orbital magnetic quantum number, with allowed values Σm_l, resulting in $2L + 1$ values; S, the total spin quantum number, with values between $s_1 + s_2$, $s_1 + s_2 - 1 \ldots |s_1 - s_2|$; M_S, the spin magnetic quantum number, with allowed values of Σm_s, resulting in $2S + 1$ values; J, the total angular momentum quantum number; and M_J, the resultant magnetic quantum number, with integral values between J and $-J$. The number of M_S values $(2S + 1)$ is called the *multiplicity*. In a similar fashion, the number of M_J values, which represents the number of degenerate energy states, is called the *statistical weight* $(g = 2J + 1)$.

The values of J may be specified by the use of one of two coupling schemes; Russell–Saunders (*LS*) coupling, which assumes that spin–orbit coupling is weak, as in the case of light atoms ($Z < 30$), and *jj* coupling, in which spin–orbit coupling is large, as typically observed in heavy atoms. In the former case, the permitted values of J are given by $L + S, L + S - 1 \ldots |L - S|$. The *jj* scheme involves addition of individual j values from each electron to give a J value for the atom (9). These coupling schemes, although useful for calculations, should be considered as limits, with many real atoms having properties intermediate to these limits. Due to the greater complexity of *jj* coupling, we will assume that Russell–Saunders coupling can be employed.

As an example, consider the ground-state sodium atom, with $1s^2 2s^2 2p^6 3s^1$. The nonvalence electrons are ignored, and hence we have $L = 0$ and $S = \frac{1}{2}$, so $J = \frac{1}{2}$. As a second example, consider an excited state of the sodium atom, $1s^2 2s^2 2p^6 3p^1$. Here $L = 1$, $S = \frac{1}{2}$, and hence $J = \frac{3}{2}$ or $\frac{1}{2}$.

A *term symbol* is a succinct way of expressing a quantum state of an atom, given by the following format: ${}^{2S+1}L_J$, where $2S + 1$ is the multiplicity of the state, L is the letter (S, P, D, F, G) that corresponds to the total orbital angular momentum quantum number, and J is the total angular momentum. This definition implies that the term symbol for ground-state sodium is ${}^2S_{1/2}$. For the $1s^2 2s^2 2p^6 3p^1$ excited state, there are two values of J, and hence two symbols; ${}^2P_{3/2}$ and ${}^2P_{1/2}$.

Only a percentage of the possible transitions between atomic states are observed experimentally. *Selection rules* have been derived from quantum mechanics to predict which transitions occur with a high degree of probability (*allowed transitions*) and cause intense lines. *Forbidden transitions* refer to transitions that are unlikely to occur and induce weak lines. For *LS* coupling, the following selection rules apply:

$$\Delta S = 0; \ \Delta L = 0, \ \pm 1 \text{ with } \Delta l = \pm 1; \ \Delta J = 0, \ \pm 1 \quad \text{but } J = 0 \text{ to } J = 0 \text{ is forbidden} \quad (2.7)$$

No change in overall spin is required for an allowed transition because spin is not directly affected by light. For an electronic transition to occur, a change in the orbital angular momentum of an electron is required. However, the coupling determines whether an overall change in orbital or total angular momentum is observed.

2.4.3 Atomic Spectra

An energy level diagram for sodium (Fig. 2.5) illustrates some of its allowed transitions (1,8). The states of sodium consist of singlet *S* states, doublet *P* states, and doublet *D* states. According to the selection rules, three major types of allowed transitions are present: those involving a *P* excited state and the *S* ground state, those involving excited *S* and *P* states, and those involving excited *D* and *P* states. Spectral transitions can be specified by the use of term symbols, with the higher energy term preceding the lower term. For example, the commonly observed emission of yellow light from sodium atoms is caused by two transitions:

$$^2P_{1/2} \rightarrow {}^2S_{1/2}(589.6\text{ nm—the sodium } D_1 \text{ line}) \tag{2.8}$$

$$^2P_{3/2} \rightarrow {}^2S_{1/2}(589.0\text{ nm—the sodium } D_2 \text{ line}) \tag{2.9}$$

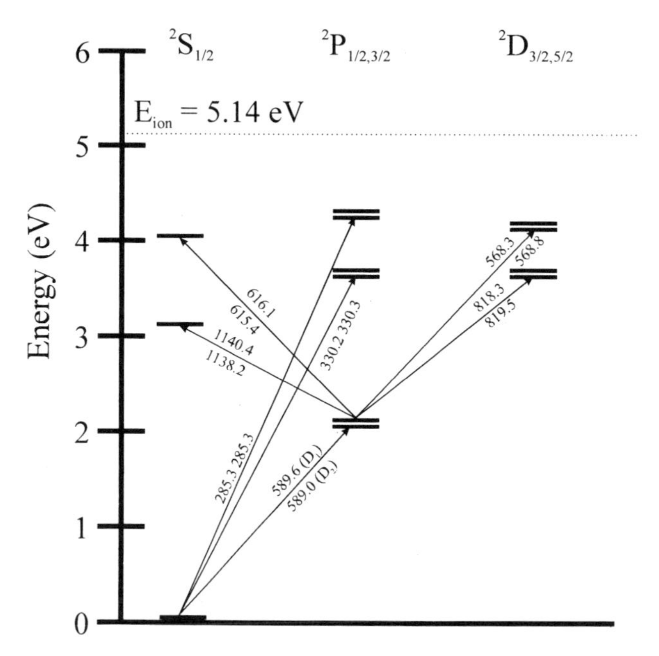

Figure 2.5. Energy level diagram for sodium: allowed transitions and their wavelengths (nm) are indicated.

The presence of two spectral lines due to differences in J values with the same value of L is called *fine structure*. The corresponding absorption lines would be written with the terms in the same order, but with the arrow reversed. Now that we have discussed the types of atomic lines present and qualitatively defined lines as allowed or forbidden, we will consider their intensities on a quantitative basis.

2.5 MEASURES OF TRANSITION STRENGTHS: TRANSITION PROBABILITIES AND OSCILLATOR STRENGTHS

The establishment of selection rules to indicate the probability of a transition is based upon evaluation of the *transition probability*, R^2 (J cm^3), by quantum mechanics for the atomic states involved in the transition (1). If the transition probability has a value of zero, then the transition is said to be forbidden. Allowed transitions have a nonzero value of R^2. The relationship between the transition probability and the Einstein coefficient for absorption for a transition between states 0 and 1 is given by:

$$B_{01} = \frac{8\pi R^2}{3h^2 g_0} \tag{2.10}$$

where h is Planck's constant and g_0 is the statistical weight of the ground state. An approximate expression for R^2 using classical physics is commonly used because of the greater ease of calculation:

$$R^2 \text{ (classical)} = \frac{3e^2 h \lambda_m g_0}{32\pi^3 \varepsilon_0 m_e c} \tag{2.11}$$

where e is the charge of the electron (1.6021×10^{-19} C); λ_m is the wavelength of maximum absorption (or emission); ε_0 is the permittivity of free space ($8.8854 \times 10^{-12}\,\mathrm{C^2/N\,m^2}$); m_e is the mass of an electron at rest (9.1094×10^{-31} kg); and c is the speed of light.

The *oscillator strength* f (dimensionless), which is another way to express transition strength, is the ratio of an experimental or theoretical measurement of the transition probability divided by the classical transition probability. The absorption oscillator strength, f_{01}, is related to the classical Einstein coefficient by:

$$f_{01} = \frac{2.50 \times 10^{-34} B_{01}}{\lambda_m} \tag{2.12}$$

Table 2.2 Oscillator Strengths of Selected Elements and Transitions in Atomic Absorption Spectrometry (77)

Element	Wavelength, nm	Oscillator Strength (f)	Accuracy[a]
Ag	328.1	0.46	B
Au	242.8	0.29	C
Ca	422.7	1.75	B
Cd	228.8	1.3	B
Co	303.4	0.023	C
Cr	357.9	0.3	C
Cu	324.7	0.32	C
Mg	285.2	1.81	B
Mn	279.5	0.57	B
Na	589.0	0.655	A
	589.6	0.327	A
Sn	286.3	0.19	B
Tl	276.8	0.30	B
Zn	213.9	1.5	B

[a] A, uncertainties <3%; B, uncertainties <10%; C, uncertainties <25%.

Values of f_{01} for several elements are listed in Table 2.2. For a strong transition, the absorption oscillator strength is typically near 1, although it may exceed that value if degenerate levels are present. The oscillator strength of weak transitions approaches 0. The absorption oscillator strength can be related to the emission oscillator strength (f_{10}) by:

$$f_{10} = f_{01} \frac{g_0}{g_1} \tag{2.13}$$

Hannaford (10) investigated the role of the oscillator strength for atomic absorption. Recent measurements of oscillator strengths were summarized. These data were employed to calculate GFAAS theoretical characteristic mass values (Section 3.2) for several elements. Doidge (11) summarized literature data for oscillator strengths of 65 elements. The implications of these data on GFAAS were discussed.

2.6 SPECTRAL PROFILES OF ABSORPTION LINES

With the resolution of detection systems typically employed in GFAAS, absorption spectra are composed of a series of narrow lines (1,4,12). However, examination with a high-resolution detection system reveals that most atomic lines are composed of two or more components, which are caused by one of two mechanisms: nuclear spin splitting and isotope shifts. The splitting of an atomic line due to an interaction between a nucleus with a nonzero orbital angular momentum and electrons is called *hyperfine structure*. For example, each sodium D line is split into two hyperfine components separated by 0.002 nm. Alternatively, the presence of more than one isotope of an element induces *isotope shifts*. All elements routinely determined by AAS, with the exceptions of calcium and silicon, have resonance lines that are affected by one or both of these broadening processes.

The finite width of each hyperfine component may be evaluated as well and explained by the use of broadening mechanisms. Broadening occurs because states 0 and 1 have finite lifetimes (Δt) according to the Heisenberg uncertainty principle,

$$E\,\Delta t \sim \frac{h}{2\pi} \tag{2.14}$$

where E is the energy of the transition and h is Planck's constant. Hence an uncertainty in Δt induces a distribution of energy levels in the atomic levels and in the wavelength distribution of light absorbed or emitted. The width of an atomic line in wavelength units ($\Delta\lambda_1$) due to these processes is given by:

$$\Delta\lambda_1 = \frac{\lambda_m^2}{c}\left[\frac{A_{10} + k_{10}}{2\pi}\right] \tag{2.15}$$

where λ_m is the wavelength of maximum absorption (or emission); c is the speed of light (2.98×10^8 m/s); A_{10} is the Einstein coefficient for spontaneous emission; and k_{10} is the rate constant (s^{-1}) for collisional deactivation from the excited state to the ground state (Fig. 2.4). The first term in the brackets represents the width of an atomic line due to *natural broadening* ($\Delta\lambda_N$):

$$\Delta\lambda_N = 5.3 \times 10^{-10}\lambda_m^2 A_{10} \tag{2.16}$$

For the sodium D transitions, the natural linewidth is on the order of 2×10^{-5} nm. In general, the contribution of the natural broadening to the linewidth is negligible compared to other broadening mechanisms.

The second term in brackets of Eq. (2.15) represents the effect of collisions between the atoms and other gaseous species upon the linewidth. Collisional broadening increases with the concentration (pressure) of potential perturbing species and is consequently also called pressure broadening. The linewidth due to collisional broadening is given by:

$$\Delta\lambda_c = \frac{\lambda_m^2}{c}\left[\left(\frac{8kT}{\pi^3}\right)^{1/2}\left(\frac{1}{M_A}+\frac{1}{\overline{M_P}}\right)^{1/2}\right]\sigma n_p \tag{2.17}$$

where c is the speed of light (2.98×10^8 m/s); k is the Boltzmann constant; T is the absolute temperature; M_A is the atomic mass of the analyte (g); $\overline{M_P}$ is the average atomic or molecular mass of the perturbing species (g); σ is the *collisional cross section* (cm^2), which represents the area occupied by the collision partners; and n_p is the density of the perturbers (cm^{-3}). In general, the collisional linewidths in commercial graphite furnaces are typically between 0.001 and 0.010 nm.

A third broadening mechanism is caused by the Doppler effect. An atom moving toward the detector displays an increase in frequency (decrease in wavelength) of its absorption transitions compared to an atom that is not moving toward the detector. Atoms moving away from the detector demonstrate a decrease in the frequency of their transitions. The width of an atomic line due to Doppler broadening ($\Delta\lambda_D$) is given by:

$$\Delta\lambda_D = 7.16 \times 10^{-7}\sqrt{\frac{T}{M_A}}(\lambda_m) \tag{2.18}$$

where T is the absolute temperature (K). Equation (2.18) demonstrates that atomic linewidths increase with temperature of the atom cell. The Doppler width is typically of the same magnitude as collisional broadening, with values between 0.001 and 0.01 nm.

The overall width of an atomic line is given by a convolution of the collisional and Doppler profiles called a *Voigt profile.* The Voigt width is given by:

$$\Delta\lambda_v \approx \left(\frac{\Delta\lambda_C}{2}\right) + \left[\left(\frac{\Delta\lambda_C}{2}\right)^2 + (\Delta\lambda_D)^2\right]^{1/2} \tag{2.19}$$

2.7 GFAAS ANALYTICAL SIGNAL: ABSORBANCE

The fundamentals of a quantitative atomic absorption measurement (1,13,14) are illustrated in Figure 2.6. In the ideal case, a monochromatic light beam from the source of intensity (I_0) enters a cell, which may contain gaseous analyte. The transmitted beam (I) then passes into a detection system that converts the light

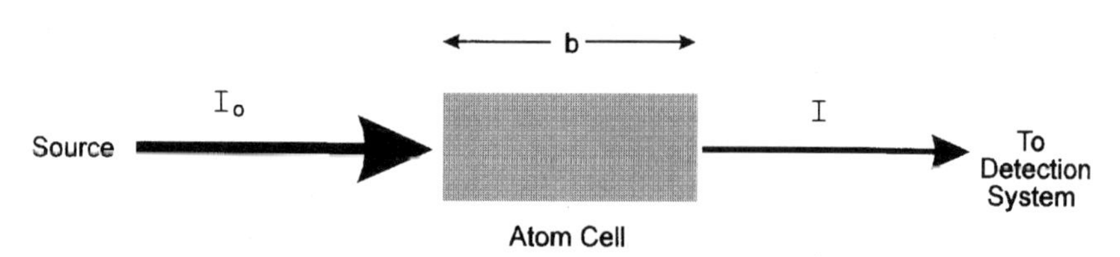

Figure 2.6. Schematic diagram of atomic absorption process, where b is the pathlength of the atom cell, I_0 is the incident intensity, and I is the transmitted intensity.

beam into an electrical signal. If analyte is present in the cell, then the transmitted beam is less intense than the incident beam. On the other hand, if no analyte is present in the cell, then the incident and transmitted beams are equal. The ratio of the transmitted beam to the incident beam is defined as the *transmittance*, T (unitless):

$$T = \frac{I}{I_0} \tag{2.20}$$

which represents the fraction of light transmitted through the cell. Alternatively, the *percent transmittance*, $\%T$, is defined as

$$T = \frac{I}{I_0} \times 100\% \tag{2.21}$$

Transmittance values range from 0, in which case no light passes through the cell, meaning a very high concentration of analyte atoms are present in the cell, to 1, in which no atoms are present in the cell. There is not a linear relationship between transmittance and concentration, and hence quantitative measurements are usually made using *absorbance*, A (unitless):

$$A = -\log T = -\log \frac{I}{I_0} = \log \frac{I_0}{I} \tag{2.22}$$

Notice that absorbance increases as transmittance decreases, indicating as more atoms are present in the cell, the absorbance increases. Absorbance is a unitless number, typically, 0 to 2, with optimum precision between 0.1 and 0.5. The quantitative relationship between absorbance and concentration (c, g/L solution) is described by the *Beer–Lambert law*:

$$A = abc \tag{2.23}$$

where a is a constant called the absorptivity (L/g cm) and b is the pathlength of the cell (cm). The magnitude and units of absorptivity are determined by the units for the pathlength and concentration. The absorptivity and concentration are related to the absorption coefficient k (cm^{-1}) by:

$$a = 0.434\frac{k}{c} \tag{2.24}$$

Atom cells for AAS typically employ a relatively long illuminated volume because of the direct proportionality between absorbance and pathlength.

2.8 INTENSITY OF ABSORBANCE IN TERMS OF FUNDAMENTAL PARAMETERS

The intensity of an absorption transition and the size of the absorption signal are dependent on whether the spectral width of the source profile is smaller (a line source) or wider (a continuum source) than the spectral profile of the atoms (Fig. 2.7) (1). We will focus on line source excitation, which has been utilized in commercial instruments. For a transition between states 0 and 1 of an atom, the absorbance is given by:

$$A = \frac{3.83 \times 10^{-13} f_{01} b n_0(t)}{\Delta\lambda_{\mathrm{eff}}} \tag{2.25}$$

where f_{01} is the oscillator strength of the transition between states 0 and 1, b is the pathlength, $\Delta\lambda_{\mathrm{eff}}$ is the width of a rectangular absorption profile, which has the same area and peak value as the light source, and $n_0(t)$ is the population of atoms in state 0 at time t. A linear relationship is predicted between concentration and absorbance, but experimentally nonlinearity occurs at relatively high concentrations because no source emits truly monochromatic light. In addition, background correction methods also reduce the linear range of the calibration graph (Section 4.6). The design and operation of sources for GFAAS are described in Section 4.1.

The GFAAS absorption signal differs from that of flame AAS because of the temporal variation of the atom population, producing a transient signal. A discrete amount of sample (10–50 μL) is introduced into the graphite furnace, which is heated to a series of temperatures for specified times. This process is called an atomization cycle (Section 4.2.2). During the atomization step, the tube is heated to a sufficiently high temperature to convert the sample into gaseous atoms. A plot of absorbance versus time shows no signal at relatively low temperatures; before the atoms have been produced, an increase in absorbance as atoms are formed, and a decrease in absorbance as atoms are

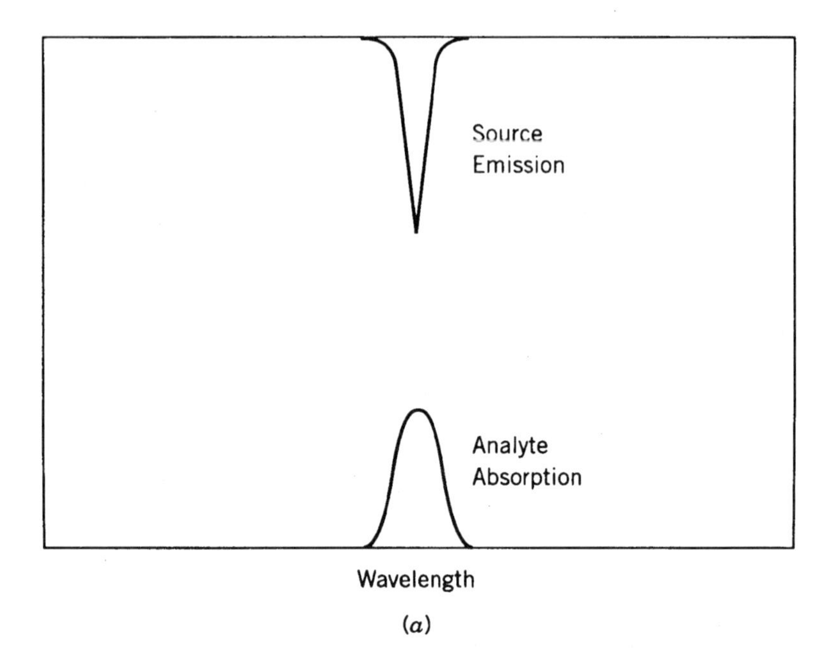

(*a*)

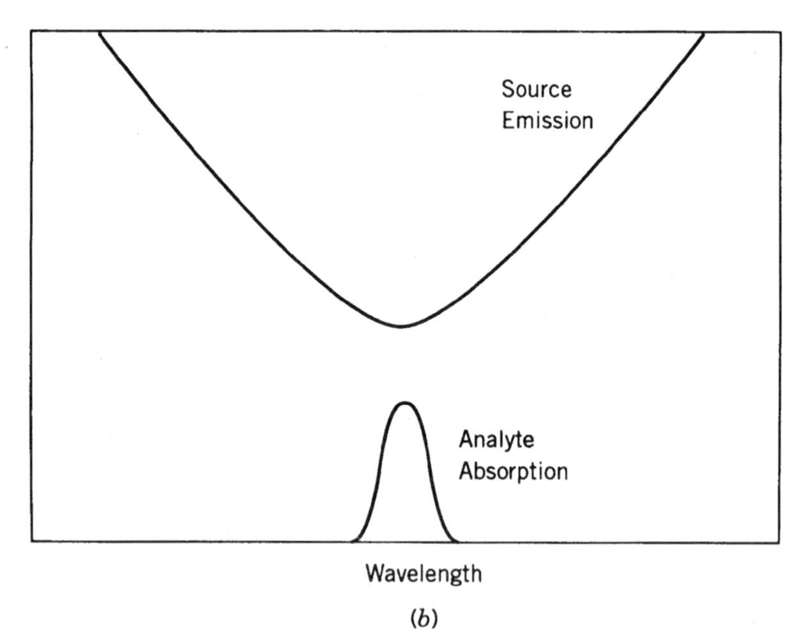

(*b*)

Figure 2.7. Spectral profile of source and atomic absorption line for (*a*) line source and (*b*) continuum source excitation.

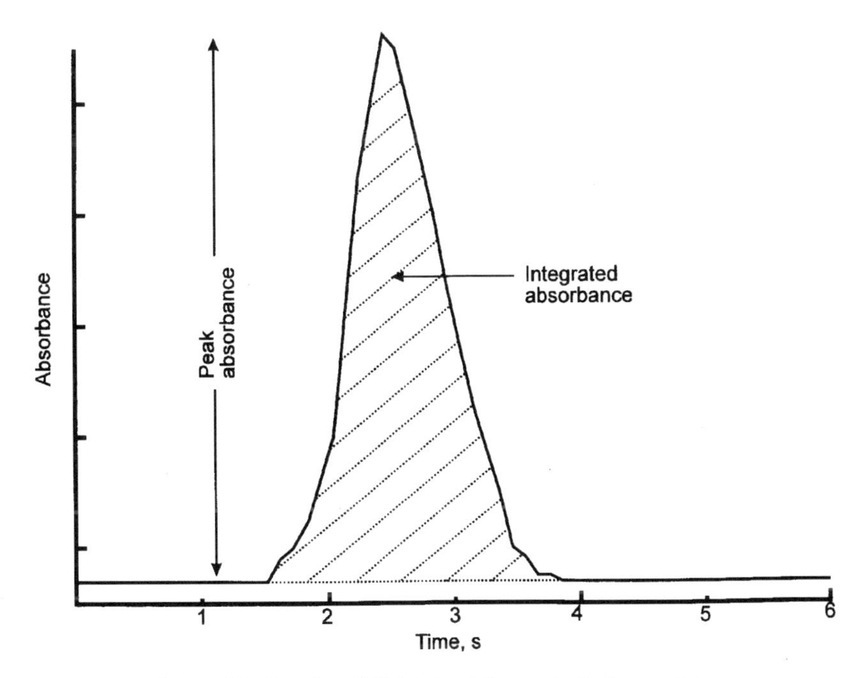

Figure 2.8. Transient GFAAS signal from a single furnace firing.

swept out of the atom cell (Fig. 2.8). The factors that determine the shape of the GFAAS signal are discussed in the following section.

2.9 NATURE OF THE TRANSIENT GFAAS SIGNAL: MECHANISM OF ATOM FORMATION IN A GRAPHITE FURNACE

Within a few years after the development of GFAAS as an analytical technique in the late 1950s and 1960s, questions were raised regarding the chemical and physical processes involved in the conversion of a (commonly) aqueous sample into gaseous atoms, followed by their removal from the furnace. Although a considerable amount of information has been obtained regarding these processes, a commercial graphite furnace is a sufficiently complex system that uncertainty exists regarding the mechanism of atom formation for many elements (15,16).

A very general mechanism for atom formation is:

$$\text{Metal salt (aq)} \xrightarrow{(1)} \text{metal salt (s)} \xrightarrow{(2)} \text{metal oxide (s)} \xrightarrow{(3)} \text{metal (s)} \xleftrightarrow{(4)} \text{metal (g)} \tag{2.26}$$

Details regarding the instrumentation are given in Chapter 4, but a brief description of the conditions is provided here. A graphite tube is typically 20 to 30 mm in length and 3 to 6 mm in diameter. The tube is surrounded by argon to prevent combustion in air at elevated temperatures. The sample is introduced into the furnace through a dosing hole (1–3 mm in diameter). In many cases, chemical compounds called chemical modifiers are also added to improve the sensitivity or accuracy for a given analyte (Section 5.1). The temperature of the tube can be controlled from ambient up to approximately 2700°C, with heating rates up 1500 °C/s. It has been shown to be beneficial for many elements to insert a platform into the tube onto which the sample is placed.

In our general mechanism [Eq. (2.26)], we have chosen to show vaporization of the metal as a starting point for the mechanisms, although some metals may vaporize as a compound. The first step involves the relatively straightforward removal of solvent (usually water) from the sample. The remaining steps include chemical/physical surface processes, such as homogeneous or heterogeneous solid–solid interactions, solid-phase nucleation, and diffusion into graphite; heterogeneous gas–solid interactions, that is, adsorption/desorption and reaction of molecules with the wall to form atoms; homogeneous gas-phase reactions; and processes by which analyte leaves the furnace. Each of these processes are discussed below, as well as the use of Monte Carlo simulation procedures to evaluate the relative rates of these processes.

2.9.1 Surface Processes

Steps (2) and (3) in Eq. (2.26) involve the chemical and physical processes that occur on the surface of the graphite tube. These processes are probably the least well understood because of the unavailability of a technique to monitor species on the surface of a commercial tube during a heating cycle. In fact, many of the techniques employ graphite substrates different from those used in GFAAS, or involve measurements made at room temperature. In spite of their limitations, these methods have revealed significant information regarding the mechanism of atom formation.

Perhaps the most widely used technique to study GFAAS surface interactions is scanning electron microscopy (SEM) (17–20), which has been used to investigate the surface morphology of graphite tubes and platforms. Welz et al. (17) performed a series of studies that elucidated the microgranularity of polycrystalline tubes and the presence of a layered structure of pyrolytically coated tubes and solid pyrolytic platforms (Section 4.2.1). Scanning electron microscopy has also been used to consider the effects of corrosive materials upon graphite substrates (17) and to visualize droplets of chemical modifiers on graphite in order to elucidate the mechanism of the modifier action.

Holcombe and co-workers (21) employed secondary ion mass spectrometry (SIMS) to investigate chemical reactions of cadmium and silver with phosphate chemical modifiers at temperatures up to 500 °C. Cadmium was shown to react with phosphate to produce a thermally stable compound that allowed the use of a higher pyrolysis temperature, although similar reactions involving silver and phosphate were not observed. A more recent SIMS study (22) investigated the presence of surface aggregates of silver and copper on graphite. Copper was shown to be dispersed across the surface at lower concentrations but to exist as microdroplets at higher concentrations. Aggregates of silver were observed at all concentrations investigated.

Majidi et al. (23–27) used Rutherford backscattering spectrometry (RBS) to study high-temperature reactions of metals with graphite. Significant migration of several metals ($\geq 3\,\mu m$) into pyrolytically coated graphite was observed. The mechanism of palladium as a chemical modifier for selenium determination was investigated. Temperature-dependent diffusion of both elements into the substrate was observed, and the lower volatility of selenium in the presence of palladium was attributed to a palladium–oxygen–selenium compound.

Vandervoort et al. (28) employed scanning tunneling microscopy to elucidate the submicrometer defect structures on pristine GFAAS graphite substrates. Polycrystalline tubes were characterized by disordered surfaces with extensive oxidation, which would be expected based on the reactivity of this material to some metals. Pyrolytically coated tubes and platforms had approximately equal distributions of relatively rough regions and microstructure and smoothly varying contours. In the latter areas, regular atomic spacing was observed between the planes with moderate curvature perpendicular to the graphite planes. The authors suggested that the good order between the layers may be responsible for the low reactivity of pyrolytically coated graphite toward metals.

Other techniques employed to characterize the surface processes include X-ray diffraction and X-ray photoelectron spectrometry (XPS) (29), gas chromatography (30), ion chromatography (31), transmission electron microscopy (32), laser desorption time-of-flight mass spectrometry (26), differential scanning calorimetry (26), thermogravimetric analysis (26), atomic force microscopy (33), and the use of radiotracers (34). We expect that other techniques will be developed to study the surface processes that occur in GFAAS.

2.9.2 Heterogeneous Solid–Gas Interactions

A number of studies have considered the adsorption/desorption processes at the graphite surface (16,35,36). These processes have been investigated by a desorption equation:

$$k = -\nu \sigma_s^n e^{-E_a/\mathcal{R}T} \tag{2.27}$$

where k is a rate constant; σ_s is the coverage of the graphite surface by the analyte (cm^{-2}); v is a constant; $\mathscr{R}$ is the gas constant (J/K mol); T is the absolute temperature (K); E_a is the activation energy (J/mol); and n is the order of release ($n \geq 0$). For $n = 0$, there is no dependence on the surface coverage of the analyte, and hence no interaction is assumed to occur between the analyte and graphite. At $n \geq 0$, the rate of desorption depends on the surface coverage, implying that attractive forces exist between the metal and graphite. Arrhenius plots have been derived from GFAAS data in order to calculate the order of release and activation energy (37–40). Accurate results are obtained by this procedure only if desorption processes dominate over gas-phase dissociation processes (41).

Heterogeneous reactions may also occur between gas-phase molecules and graphite to produce gaseous atoms. Two-dimensional spatial and temporal images of atoms and molecules in a graphite tube have demonstrated the importance of these reactions for some elements. The earlier work, which employed a cine camera as a detector, was called shadow spectral filming (SSF) (42–46); more recent work used a charge-coupled device (Section 4.4) and was called shadow spectral digital imaging (SSDI) (47–51). For example, for aluminum, the highest concentration of aluminum atoms was adjacent to the walls, but the highest density of aluminum-containing molecules was in the central axis of the tube (47) (Fig. 2.9). The authors concluded that the

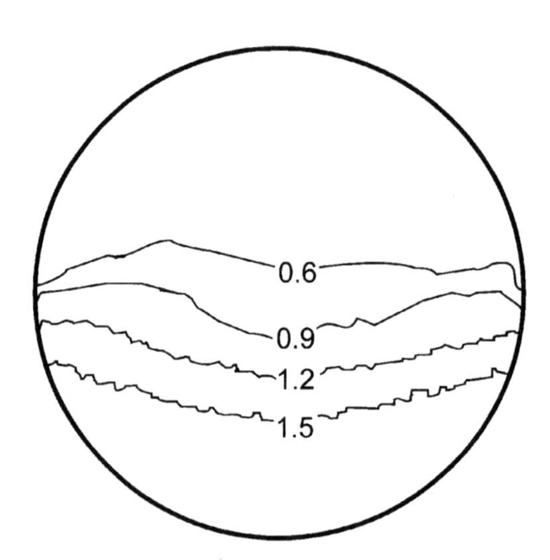

Figure 2.9. Absorbance contour map obtained by SSDI for vaporization of aluminum-containing molecules (47).

mechanism of atomization included the following heterogeneous reduction reaction:

$$Al_xO_{(g)} + C_{(S)} \rightarrow xAl_{(g)} + CO_{(g)} \tag{2.28}$$

where $x = 1, 2$. Recently, Sturgeon and co-workers (52) used this technique to observe transient atomization events in a furnace atomization plasma emission spectrometric (FAPES) source (Section 4.7).

2.9.3 Gas-Phase Chemistry

A considerable amount of information has been obtained regarding identity and the temporal and spatial distributions of atoms and molecules in the gas phase of a graphite tube. The shadow imaging techniques described above have provided valuable quantitative information about the distribution of species. The identification of species in a graphite tube has involved the use of mass spectrometry (MS) (16,53–63). In one scheme, MS methods have employed operation of the tube under vacuum conditions, which is relatively easy to set up but has the disadvantage of significantly reducing gas-phase collisions compared to "conventional" operation at atmospheric pressure. Alternatively, the tube is operated at atmospheric pressure, and a series of vacuum pumping stages are used to reduce the pressure to the mass spectrometer. This system allows gas-phase reactions to occur in the furnace, although additional reactions may take place as the pressure is reduced. A combination of these conditions has been used to distinguish between condensed-phase and gas-phase interactions; the latter would be expected to be negligible using vacuum atomization, while the former would be unaffected. An example of the mass spectral data obtained is shown in Figure 2.10 (53).

Mass spectrometry data have been used to elucidate the mechanism of a variety of elements (16). We will consider beryllium as an example element (59). The atomization of beryllium nitrate was accompanied by the presence of beryllium monoxide, $(BeO)_2$ (monoxide dimer), $(BeO)_4$ (monoxide tetramer), and several carbides (Be_2C, BeC_2, Be_2C_2, and Be_2C_4). The authors concluded that the primary source of gaseous beryllium was the dissociation of adsorbed beryllium monoxide:

$$BeO_{(ad)} \rightarrow Be_{(g)} + O_{(g)} \tag{2.29}$$

Although a tremendous amount of information has been deduced from mass spectral data, considerable uncertainty and disagreement still exists in the literature due to the absence of "real-time" data regarding processes at the graphite surface. Perhaps the most studied element is aluminum, but questions

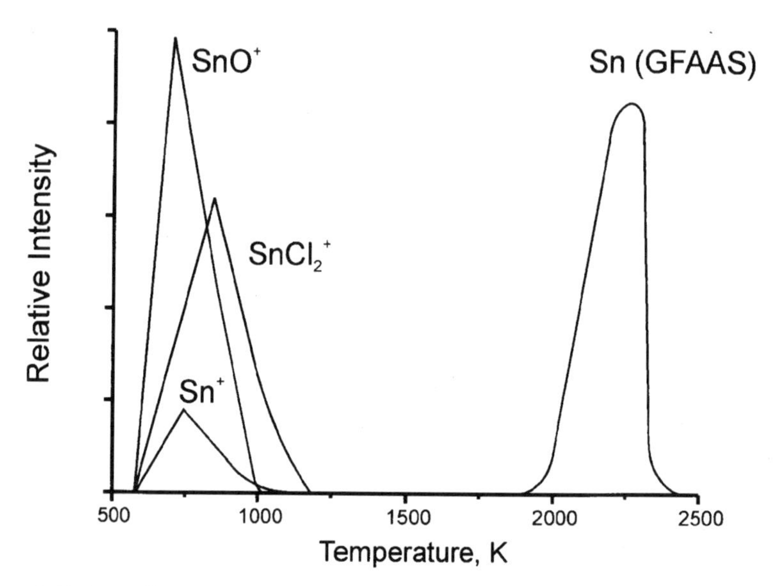

Figure 2.10. Normalized temperature profiles of GFAAS and mass spectral signals for atmospheric vaporization of $SnCl_2$ (200 ng Sn, 0.8% HCl) from a pyrolytically coated graphite tube. Electron energy: 70 eV; absorption at 286.3 nm; and *m/z* values for Sn^+, SnO^+, and $SnCl_2^+$ were 120, 136, and 190, respectively (53).

remain today regarding the mechanism of atomization for this element (61,64,65). For this reason, mechanistic investigations continue to be an active area of research in GFAAS.

2.9.4 Processes by Which Analyte Is Lost from the Tube

The principal loss mechanisms of analyte from a graphite tube are diffusion, thermal expansion, and expulsion caused by gas evolution from a matrix (66). The average residence time of an analyte (t_d) (13) is given by:

$$t_d = \frac{(l_t)^2}{8D_T} \tag{2.30}$$

where l_t is the length of the graphite tube (cm) and D_T is the diffusion coefficient of the analyte in argon at temperature T (cm^2/s). This simple expression does not consider diffusion through the dosing hole, through which a significant fraction of atoms may be lost (Section 2.9.5). The diffusion coefficient at any temperature may be calculated from the following relationship (67,68):

$$D_T = D_{273}\left(\frac{T}{273}\right)^{n_d} \tag{2.31}$$

where D_{273} is the diffusion coefficient at 273 K and n_d is an exponential factor between 1.65 and 1.96.

Thermal expansion (66) may be quantified by the time t_T required to remove a fraction of the tube volume $(x_0/x - 1)$ by fluid flow considerations:

$$t_T = \frac{T_0}{\beta}\left(\frac{x_0}{x} - 1\right) \tag{2.32}$$

where x_0 is the distance between the center and the end of the tube (cm); x is a position on the tube (cm); T_0 is the temperature at time $T=0$ (before atomization) (K); and β is the heating rate (K/s).

The effect of large quantities of gas due to decomposition or vaporization of a matrix upon the analytical signal may be expressed by the movement of vapors a distance ∂x by:

$$\partial x = \frac{\partial Q\, RT}{PA} \tag{2.33}$$

where ∂Q is the number of moles of gas injected into the tube; R is the ideal-gas constant (8.314 J/K mol); T is the absolute temperature (K); P is the pressure (normally assumed to be 1 atm); and A is the area of the furnace (cm^2).

Holcombe (66) evaluated the relative importance of each of these loss mechanisms. Diffusion is most significant at low heating rates and high atomization temperatures, while thermal expansion contributes most strongly for volatile elements with a high heating rate. Thermal expulsion by the matrix is only significant if more than 10^{-7} moles of gas are evolved, and if the vaporization of matrix and analyte are nearly coincident.

2.9.5 Monte Carlo Simulations of Atomization Processes

Holcombe and co-workers (69–76) have investigated the use of Monte Carlo techniques in order to simulate processes that occur in a graphite furnace. This approach involves the use of random numbers with probabilities obtained from macroscopic properties (72). Although computationally intensive, Monte Carlo methods often provide an easier route to the solution of complex problems, such as modeling graphite furnace processes. Histen et al. (76) recently published a Monte Carlo program for use on a 486 or higher personal computer.

The chemical and physical processes involved in Monte Carlo simulations are given by

$$M_{(c)} \xrightarrow{k_1} M_{(g)} \underset{k_3}{\overset{k_2}{\rightleftarrows}} M'_{(c)} \tag{2.34}$$

where $M_{(c)}$ and $M'_{(c)}$ are analyte in condensed phases; $M_{(g)}$ is analyte in the gas phase; and k_1, k_2, and k_3 are rate constants that may be evaluated by Eq. (2.27). A number of geometric variables may be specified that include the length and diameter of the tube, diameter and position of deposition, the position and diameter of the dosing hole, the number of atoms present, and the temperature of the tube. This approach has been employed to investigate mechanisms of atomization for a variety of elements. The availability of this program for personal computers provides a convenient way to simulate GFAAS atomization processes (76).

REFERENCES

1. J. D. Ingle and S. R. Crouch, *Spectrochemical Analysis*, Prentice-Hall, Englewood Cliffs, NJ, 1988.
2. D. A. Skoog, D. M. West, and F. J. Holler, *Fundamentals of Analytical Chemistry*, 7th ed, Saunders College Publishing, Fort Worth, TX, 1996.
3. C. T. J. Alkemade and R. Herrmann, *Fundamentals of Analytical Flame Spectroscopy*, Wiley, New York, 1979.
4. D. J. Butcher, J. P. Dougherty, J. T. McCaffrey, F. R. Preli, A. P. Walton, and R. G. Michel, *Prog. Anal. Spectrosc.*, **10**, 359 (1987).
5. A. C. G. Mitchell and M. W. Zemansky, *Resonance Radiation and Excited Atoms*, Macmillan, Cambridge, UK, 1934.
6. G. Herzberg, *Atomic Spectra and Atomic Structure*, Dover, New York, 1944.
7. I. I. Sobelman, *Atomic Spectra and Radiative Transitions*, 2nd ed, Springer, Berlin, 1992.
8. P. W. Atkins, *Physical Chemistry*, W. H. Freeman, New York, 1990.
9. C. W. Haigh, *J. Chem. Ed.*, **72**, 206 (1995).
10. P. Hannaford, *Spectrochim. Acta, Part B*, **49B**, 1581 (1994).
11. P. S. Doidge, *Spectrochim. Acta, Part B*, **50B**, 209 (1995).
12. V. B. E. Thomsen, *J. Chem. Ed.*, **72**, 616 (1995).
13. B. Welz, *Atomic Absorption Spectrometry*, 2nd ed, VCH, Weinheim, Germany, 1985.

14. D. A. Skoog and J. J. Leary, *Principles of Instrumental Analysis*, 4th ed, Saunders, Fort Worth, TX, 1992.
15. B. V. L'vov, *Spectrochim. Acta, Part B*, **52B**, 1 (1997).
16. D. L. Styris and D. A. Redfield, *Spectrochim. Acta Rev.*, **15**, 71 (1993).
17. B. Welz, G. Schlemmer, H. M. Ortner, and W. Wegscheider, *Prog. Anal. Spectrosc.*, **12**, 111 (1989).
18. H. Qiao and K. W. Jackson, *Spectrochim. Acta, Part B*, **46B**, 1841 (1991).
19. H. Qiao and K. W. Jackson, *Spectrochim. Acta, Part B*, **47B**, 1267 (1992).
20. T. M. Mahmood, H. Qiao, and K. W. Jackson, *J. Anal. Atom. Spectrom*, **10**, 43 (1995).
21. D. C. Hassell, V. Majidi, and J. A. Holcombe, *J. Anal. Atom. Spectrom.*, **6**, 105 (1991).
22. J. G. Jackson, R. W. Fonesca, and J. A. Holcombe, *J. Anal. Atom. Spectrom.*, **9**, 167 (1994).
23. C. Eloi, J. D. Robertson, and V. Majidi, *J. Anal. Atom. Spectrom.*, **8**, 217 (1993).
24. C. C. Eloi, J. D. Robertson, and V. Majidi, *Anal. Chem.*, **67**, 335 (1995).
25. V. Majidi and J. D. Robertson, *Spectrochim. Acta, Part B*, **46B**, 1723 (1991).
26. V. Majidi, R. G. Smith, R. E. Bossio, R. T. Pogue, and M. W. McMahon, *Spectrochim. Acta, Part B*, **51B**, 941 (1996).
27. C. C. Eloi, J. D. Robertson, and V. Majidi, *Appl. Spectrosc.*, **51**, 236 (1997).
28. K. G. Vandervoort, D. J. Butcher, C. T. Brittain, and B. B. Lewis, *Appl. Spectrosc.*, **50**, 928 (1996).
29. Y. Peng-yuan, N. Zhe-ming, Z. Zhi-xia, X. Fu-chun, and J. An-bei, *J. Anal. Atom. Spectrom.*, **7**, 515 (1992).
30. R. E. Sturgeon, K. W. M. Siu, G. J. Gardiner, and S. S. Berman, *Anal. Chem.*, **58**, 42 (1986).
31. M. M. Chaudhry and D. Littlejohn, *Analyst*, **117**, 713 (1992).
32. K. Yasuda, Y. Hirano, T. Kamino, and K. Hirokawa, *Anal. Sci.*, **10**, 623 (1994).
33. J. Habicht, T. Prohaska, G. Friedbacher, M. Grasserbauer, and H. M. Ortner, *Spectrochim. Acta, Part B*, **50B**, 713 (1995).
34. A. B. Volynsky, V. Krivan, and S. V. Tikhomirov, *Spectrochim. Acta, Part B*, **51B**, 1253 (1996).
35. J. A. Holcombe and G. D. Rayson, *Prog. Anal. Atom. Spectrosc.*, **6**, 225 (1983).
36. A. B. Volynsky, *Spectrochim. Acta, Part B*, **51B**, 1573 (1996).
37. S. L. Paveri-Fontano, G. Tessari, and G. Torsi, *Anal. Chem.*, **46**, 1032 (1974).
38. G. Torsi and G. Tessari, *Anal. Chem.*, **47**, 839 (1975).
39. G. Tessari, *Anal. Chem.*, **47**, 842 (1975).
40. B. Smets, *Spectrochim. Acta, Part B*, **35B**, 33 (1980).
41. J. A. Holcombe, *Spectrochim. Acta, Part B*, **44B**, 975 (1989).
42. C. W. Huie and C. J. Curran, *Appl. Spectrosc.*, **42**, 1307 (1988).

43. C. W. Huie and C. J. Curran, *Appl. Spectrosc.*, **44**, 1329 (1990).
44. A. K. Gilmutdinov, Y. A. Zakharov, V. P. Ivanov, and A. V. Voloshin, *J. Anal. Atom. Spectrom.*, **6**, 505 (1991).
45. A. K. Gilmutdinov, Y. A. Zakharov, V. P. Ivanov, A. V. Voloshin, and K. Dittrich, *J. Anal. Atom. Spectrom.*, **7**, 675 (1992).
46. A. K. Gilmutdinov, Y. A. Zakharov, and A. V. Voloshin, *J. Anal. Atom. Spectrom.*, **8**, 387 (1993).
47. C. L. Chakrabarti, A. K. Gilmutdinov, and J. C. Hutton, *Anal. Chem.*, **65**, 716 (1993).
48. D. M. Hughes, C. L. Chakrabarti, D. M. Goltz, R. E. Sturgeon, and D. C. Grégoire, *Appl. Spectrosc.*, **50**, 715 (1996).
49. A. K. Gilmutdinov, B. Radziuk, M. Sperling, and B. Welz, *Spectrochim. Acta, Part B*, **51B**, 1023 (1996).
50. W. Frech and D. C. Baxter, *Spectrochim. Acta, Part B*, **51B**, 961 (1996).
51. D. M. Hughes, C. L. Chakrabarti, M. M. Lamoureux, J. C. Hutton, D. M. Goltz, R. E. Sturgeon, D. C. Grégoire, and A. K. Gilmutdinov, *Spectrochim. Acta, Part B*, **51B**, 973 (1996).
52. R. E. Sturgeon, V. Pavski, and C. L. Chakrabarti, *Spectrochim. Acta, Part B*, **51B**, 999 (1996).
53. G. N. Brown and D. L. Styris, *J. Anal. Atom. Spectrom*, **8**, 211 (1993).
54. L. J. Prell, D. L. Styris, and D. A. Redfield, *J. Anal. Atom. Spectrom.*, **5**, 231 (1990).
55. L. J. Prell and D. L. Styris, *Spectrochim. Acta, Part B*, **46B**, 45 (1991).
56. L. J. Prell, D. L. Styris, and D. A. Redfield, *J. Anal. Atom. Spectrom.*, **6**, 25 (1991).
57. D. L. Styris, *Anal. Chem.*, **56**, 1070 (1984).
58. D. L. Styris, *Fresenius J. Anal. Chem.*, **323**, 710 (1986).
59. D. L. Styris and D. A. Redfield, *Anal. Chem.*, **59**, 2897 (1987).
60. D. L. Styris, L. J. Prell, D. A. Redfield, J. A. Holcombe, D. A. Bass, and V. Majidi, *Anal. Chem.*, **63**, 508 (1991).
61. J. A. Holcombe, D. L. Styris, and J. D. Harris, *Spectrochim. Acta, Part B*, **46B**, 629 (1991).
62. M. S. Droessler and J. A. Holcombe, *Spectrochim. Acta, Part B*, **42B**, 981 (1987).
63. M. S. Droessler and J. A. Holcombe, *J. Anal. Atom. Spectrom.*, **2**, 785 (1987).
64. D. A. Katskov, A. M. Shtepan, I. L. Grinshtein, and A. A. Pupyshev, *Spectrochim. Acta, Part B*, **47B**, 1023 (1992).
65. B. V. L'vov, *Spectrochim. Acta, Part B*, **51B**, 533 (1996).
66. J. A. Holcombe, *Spectrochim. Acta, Part B*, **38B**, 609 (1983).
67. B. V. L'vov, *Spectrochim. Acta, Part B*, **45B**, 633 (1990).
68. J. M. Harnly and B. Radziuk, *J. Anal. Atom. Spectrom.*, **10**, 197 (1995).
69. S. S. Black, M. R. Riddle, and J. A. Holcombe, *Appl. Spectrosc.*, **40**, 925 (1986).
70. O. A. Güell and J. A. Holcombe, *Spectrochim. Acta, Part B*, **43B**, 459 (1988).
71. O. A. Güell and J. A. Holcombe, *Spectrochim. Acta, Part B*, **44B**, 185 (1989).

72. O. A. Güell and J. A. Holcombe, *Anal. Chem.*, **62**, 529A (1990).
73. O. A. Güell and J. A. Holcombe, *Appl. Spectrosc.*, **45**, 1171 (1991).
74. O. A. Güell and J. A. Holcombe, *J. Anal. Atom. Spectrom.*, **7**, 135 (1992).
75. O. A. Güell and J. A. Holcombe, *Spectrochim. Acta, Part B*, **47B**, 1535 (1992).
76. T. E. Histen, O. A. Güell, I. A. Chavez, and J. A. Holcombe, *Spectrochim. Acta, Part B*, **51B**, 1279 (1996).
77. J. W. Robinson, *CRC Practical Handbook of Spectroscopy*, CRC Press, Boca Raton, FL, 1991.

CHAPTER 3

QUANTITATIVE GFAAS: CALIBRATION

This chapter discusses methods employed to determine the concentrations of elements in samples by GFAAS. General definitions, including precision and accuracy, provide an introduction to these concepts. Parameters more specific to GFAAS are also defined and discussed, such as peak and integrated absorbance and characteristic mass. Various calibration methods are described, including the use of aqueous calibration graphs, standard additions, and internal standards. The concept of absolute, or standardless, analysis is introduced at the conclusion of the chapter.

3.1 GENERAL ANALYTICAL METHODOLOGY

Precision and accuracy are fundamental concepts employed in quantitative analysis (1–6). All measurements have random error, which can be characterized statistically. *Precision* refers to the repeatability of a measurement; a technique that provides highly reproducible results for several measurements is said to have good precision. Quantitatively the precision is usually defined in terms of the *standard deviation* s_D:

$$s_D = \sqrt{\sum \frac{(x_i - \bar{x})^2}{n-1}} \tag{3.1}$$

where x_i is an individual measurement, $\bar{x}$ is the mean of the measurements, and n is the number of measurements. The precision is usually reported by multiplying the standard deviation by a factor of 2 or 3, so that measurements are reported as $\bar{x} \pm 2s_D$ or $\bar{x} \pm 3s_D$. The relative value of the precision is usually determined by use of the *relative standard deviation* (RSD) (also called *coefficient of variation*), which is calculated by the following equation:

$$\text{RSD} = \frac{s_D}{\bar{x}} \times 100 \tag{3.2}$$

Accuracy refers to the closeness between a measured value and the "true" or accepted value. Consequently, in order to assess accuracy, it is necessary to have samples that have been previously analyzed by two or more independent methods. These *standard reference materials* (SRMs) allow characterization of errors that may be categorized as instrumental, personal, and method (see Section 7.3). Instrumental errors can usually be minimized by monitoring the operation of the spectrometer and regular calibration. Personal errors can be eliminated by self-discipline and good technique. Many sources of personal error have been minimized with the development of automated instrumentation. For GFAAS, method errors, which may involve losses of analyte in sample preparation (Chapter 6), spectral interferences (Section 5.1), or chemical interferences (Section 5.2), are usually the most difficult to eliminate. Careful method development and quality control procedures (Chapter 7) are required to minimize these errors.

In practice, available SRMs do not exactly match the components of a sample with unknown composition, and hence quality control procedures must be employed to evaluate the accuracy of analyses (Chapter 7). For example, the addition of analyte as a standard to samples, which is called a *recovery check*, may be used to assess whether chemical interferences (Section 5.2) are present. The presence of spectral interferences may be identified by dilution of the sample. If the signal is smaller than would be expected by the dilution factor, then an uncorrected background signal is present. Alternatively, the determination may be made at a second wavelength or by use of a second method of background correction. If the background has been accurately corrected for, one would expect identical values by these methods. Quality control procedures are discussed in more detail in Chapter 7.

3.2 CALIBRATION GRAPHS

The basis of quantitative absorption analysis is given by the Beer–Lambert law (1–6) (Section 2.7):

$$A = abc \tag{3.3}$$

where c is the concentration (g/L), a is a constant called the absorptivity (L/g cm), and b is the pathlength of the cell (cm). Atom cells for AAS typically employ a relatively long illuminated volume because of the direct proportionality between absorbance and pathlength.

Absorbance is a unitless number, typically 0 to 2. Graphite furnace AAS is a relative method of analysis, which means that standard solutions must be used to obtain quantitative results. A discrete amount of sample (10–50 μL) is introduced into the graphite furnace, which is heated to a series of temperatures for specified times. This process is called an atomization cycle (Section 4.2.2).

During the atomization step, the tube is heated to a sufficiently high temperature to convert the sample into gaseous atoms. A plot of absorbance versus time shows no signal at relatively low temperatures. Before the atoms have been produced, an increase in absorbance as atoms are formed, and a decrease in absorbance as atoms are swept out of the atom cell (Fig. 2.8). Quantitative evaluation of the GFAAS signal is best done by measuring the area under the transient peak, which is called *integrated absorbance* (peak area, units of seconds). Relatively early GFAAS work employed *peak absorbance* (peak height, unitless) measurements, which are less suitable because considerable variation in peak absorbance may occur depending on the sample matrix present and the age of a graphite tube. The mass of analyte, rather than concentration, is generally plotted on the x axis. Analyte mass is more appropriate for GFAAS because a discrete amount of sample is introduced into the atomizer, and hence the use of concentration is ambiguous because of its dependence on the volume introduced (usually 10–25 μL, Section 3.2.2). Typical masses introduced for GFAAS are in the picogram to low nanogram range, depending on the element, which translates to solution concentrations in the picogram/milliliter (part per trillion) to nanogram/milliliter (part per billion) levels.

Calibration is commonly achieved by preparing a graph of integrated absorbance (y axis) versus the mass of analyte introduced in standard solutions (x axis) called a *calibration graph* (1–6) or *curve*. A mathematical relationship, which is usually obtained by linear regression, is established between absorbance and concentration. The absorbance of unknowns is then measured, and the concentration of analyte is determined from this relationship. A typical calibration graph for GFAAS is shown in Figure 3.1. Generally calibration is performed using the linear portion of the analytical graph, that is, when absorbance is directly proportional to mass of analyte. The useful linear range of the graph is defined by the lowest mass [called the *level of quantitation* (LOQ)] and highest mass [called the *level of linearity* (LOL)] standard in this region. The level of quantitation is defined as the concentration at which the RSD exceeds a specified amount (e.g., 10%), while the level of linearity represents the concentration at which the slope deviates by a specified amount from the linear region. Usually the extent of linearity of a graph is described by the useful *linear dynamic range* (LDR), which is defined as

$$\mathrm{LDR} = \log\left(\frac{\mathrm{LOL}}{\mathrm{LOQ}}\right) \tag{3.4}$$

This means that if a calibration graph is linear from 10 to 1000 pg, then the linear dynamic range is defined as two orders of magnitude. Generally at least five standards should be employed to prepare a calibration curve.

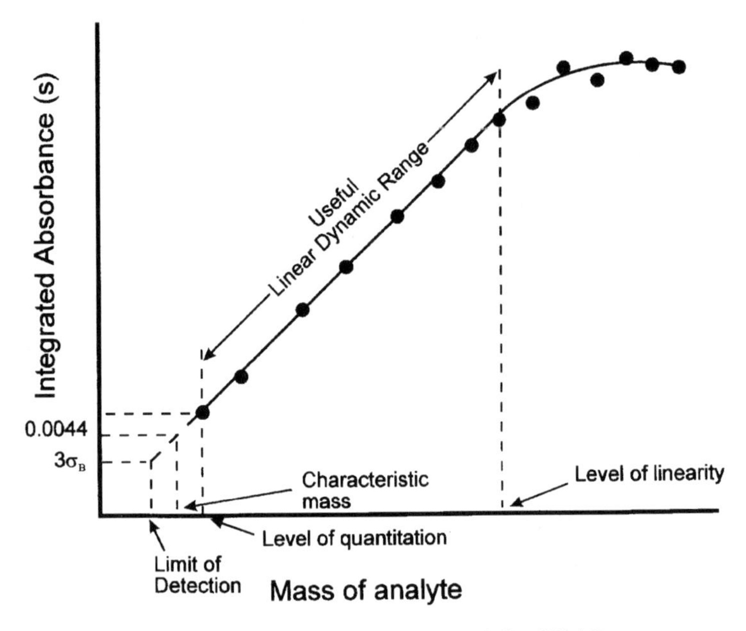

Figure 3.1. Aqueous calibration graph for GFAAS.

The slope of a calibration curve is called the *calibration sensitivity* (s). The sensitivity of a particular instrument is evaluated by use of the *characteristic mass*, m_0, which represents the mass of analyte that gives an integrated absorbance of 0.0044 s:

$$m_0 = \frac{0.0044}{s} \tag{3.5}$$

The smallest mass of analyte that can be distinguished from statistical fluctuations in a blank, which usually corresponds to the standard deviation of the blank absorbance times a constant, is called the *limit of detection* (LOD). The limit of detection is most commonly defined as the mass of analyte that gives a signal equal to three times the standard deviation on the blank (σ_B):

$$\text{LOD} = \frac{3\sigma_B}{s} \tag{3.6}$$

For most elements commonly determined by GFAAS, the characteristic mass and LOD are between 1 and 50 pg, and the level of quantitation is approximately a factor of 5 to 10 above the LOD. The GFAAS method development usually involves evaluation of the characteristic mass and LOD from higher value standards. The experimental values are compared to standard values provided by the instrument manufacturer to evaluate whether conditions have been optimized. Interferences may induce degradation of these figures of merit, although the use of modern furnace technology has eliminated these interferences for many samples. A more detailed description of limits of detection is provided by Kirchmer (1).

It should be pointed out that, with the use of a microcomputer, calibration can be routinely performed on nonlinear graphs, although the precision is usually degraded. Consequently, it is generally advisable to ensure that samples are calibrated on the linear portion. The principal cause of nonlinearity of the graphs is caused by the inability of any light source to produce truly monochromatic radiation, as specified by Beer's law. Hollow cathode lamps emit various types of stray light, which may include a relatively broad analytical emission line, or if the spectral bandpass is sufficiently wide, the transmission of a nonabsorbable line (e.g., a second analyte line or a line emitted by the fill gas) to the detector. The latter problem may occur with elements such as iron, which have a large number of emission lines, if the spectral bandpass is excessively wide. The use of background correction may also reduce the LDR (Section 4.6). Linearization procedures, which have been developed to calibrate nonlinear data, are discussed in more detail in Section 4.6.4.1.

3.3 METHOD OF STANDARD ADDITIONS

The *method of standard additions* (2,4–6) is employed to compensate for a decrease in calibration sensitivity caused by the presence of matrix components (chemical interferences, Section 5.2). This method involves the addition of various amounts of analyte (as standard) to aliquots of the sample. A graph is then prepared of mass of analyte added as standard (x axis) versus integrated absorbance (y axis). Linear regression is used to define a mathematical relationship (line) between absorbance and mass. The resulting standard additions graph is shown in Figure 3.2. The presence of a nonzero absorbance at zero added analyte is attributable to the analyte present in the sample. Extrapolation of the line to the x axis allows calculation of the amount of analyte in the sample.

In the early days of GFAAS, standard addition was widely used for calibration, and it is still used today, although its limitations must be understood. In order to obtain accurate results, all standard addition solutions must be within the linear range of the calibration curve since the use of standards that exceed the LOL will give erroneously high results. In addition, the added analyte must be affected by the interferences in an identical fashion to the analyte inherent to

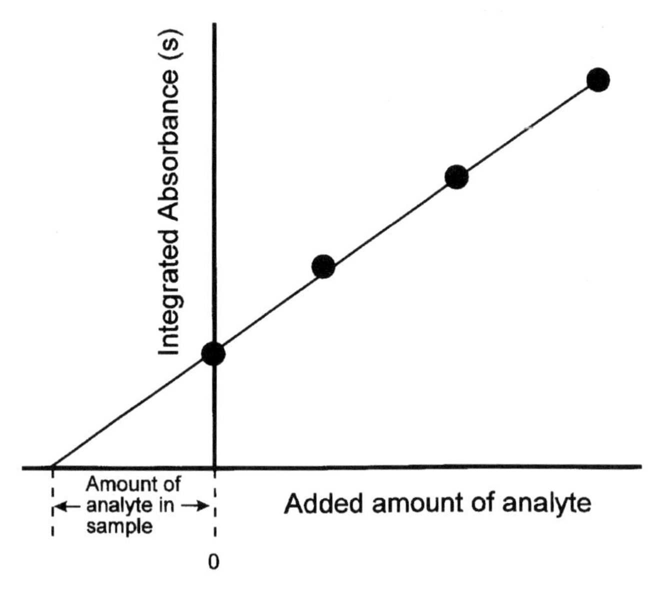

Figure 3.2. Standard additions graph for GFAAS.

the sample. Standard addition does not account for interferences caused by the absorption, emission, or scattering of light by nonanalyte (matrix) components (e.g., spectral interferences, Section 5.1). Welz (7) discussed the misuse of the method of standard addition and how this may cause inaccurate analyses. The use of modern furnace technology has reduced interferences and the application of standard addition for GFAAS.

3.4 METHOD OF INTERNAL STANDARDIZATION

The *internal standard method* (4,6) involves a comparison of the detector response of the analyte to a second element (the internal standard) that has been added to the standards and samples, or if present in the sample, has been added to the standards only. In order to use this method, the instrument must be capable of monitoring at least two elements simultaneously. The use of the internal standard method has not been widely employed in atomic absorption because of the relative scarcity of multielemental instrumentation until recently (Section 4.8.2). The usual practice is to employ one concentration of internal standard with a series of analyte standards. A graph is then prepared of the ratio of the analyte absorbance (A_a) to the internal standard absorbance (A_{is}), A_a/A_{is}, versus the mass of analyte.

A suitable internal standard has similar chemical and spectral properties to the analyte, and hence this technique ideally will compensate for a change in atomizer conditions or the presence of spectral interferences. Internal standardization has the advantage of providing improved precision because of these corrections. This technique may also reduce the effects of chemical interferences if both the analyte and internal standard are affected equally by an interferent. If the internal standard is added prior to sample preparation procedures, and the analyte and internal standard behave identically, losses in sample treatment may be compensated for.

Unfortunately, this technique is only applicable if a suitable internal standard is available, which must have similar properties to the analyte, be determinable by GFAAS, be soluble in the samples, and not interfere with the analyte determination. With the increased number of multielement commercial atomic absorption spectrometers (Section 4.8.2), the method of internal standardization will probably find increased use in GFAAS.

3.5 ABSOLUTE ANALYSIS

Absolute analysis (8–12) involves doing quantitative analyses from a theoretical equation and a single measurement of a sample. Absolute, or standardless, methods, such as neutron activation analysis, consequently do not require the preparation of calibration graphs prior to analysis. The development of standardless GFAAS has been a goal of researchers in recent years (8–12). Torsi and co-workers (10–13) constructed an electrothermal atomizer that has been used for absolute analysis using a modification of the Beer–Lambert law:

$$A_{\mathrm{peak}} = \frac{KN_0}{\sigma} \tag{3.7}$$

where A_{peak} is the peak absorbance of a GFAAS measurement; N_0 is the number of atoms introduced into the atom cell (atoms); σ is the collisional cross section of the atom cell (cm^2); and K is a spectroscopic constant (cm^2/atom). Absolute measurements require a furnace system that is capable of simultaneously vaporizing all the atoms introduced into the furnace in order to reliably measure A_{peak}. Under these conditions, it is possible to measure K with good accuracy and precision. For example, Locatelli and co-workers (12) reported a value of K of $(1.08 \pm 0.07) \times 10^{13}\ \mathrm{cm}^2$/atom. This furnace system has been combined with an electrostatic precipitation method for the determination of lead in aerosols (11,12) (Section 6.2). Standardless analysis is particularly useful for these samples because of the absence of aerosol standard reference materials.

Although absolute analysis has considerable potential for analysis, currently laboratory-constructed furnace systems are required to obtain results of high precision and accuracy. We therefore recommend that most analysts use one of the calibration methods described in Sections 3.2 to 3.4.

REFERENCES

1. C. J. Kirchmer, "Limits of Detection and Accuracy in Trace Elements Analysis," in Z. B. Alfassi, Ed., *Determination of Trace Elements*, VCH, Weinheim, Germany, 1994.
2. D. A. Skoog and J. J. Leary, *Principles of Instrumental Analysis*, 4th ed, Saunders, Fort Worth, TX, 1992.
3. H. A. Strobel and W. R. Heineman, *Chemical Instrumentation: A Systematic Approach*, 3rd ed., Wiley, New York, 1989.
4. C. Vandecasteele and C. B. Block, *Modern Methods for Trace Element Determination*, Wiley, New York, 1993.
5. B. Welz, *Atomic Absorption Spectrometry*, 2nd ed., VCH, Weinheim, Germany, 1985.
6. J. D. Ingle and S. R. Crouch, *Spectrochemical Analysis*, Prentice-Hall, Englewood Cliffs, NJ, 1988.
7. B. Welz, *Fresenius J. Anal. Chem.*, **325**, 95 (1986).
8. W. Slavin and G. R. Carnrick, *Spectrochim. Acta, Part B*, **39B**, 271 (1984).
9. B. V. L'vov, *Spectrochim. Acta, Part B*, **45B**, 633 (1990).
10. G. Torsi, P. Reschiglian, F. Fagioli, and C. Locatelli, *Spectrochim. Acta, Part B*, **48B**, 681 (1993).
11. G. Torsi, P. Reschiglian, M. T. Lippolis, and A. Toschi, *Microchem. J.*, **53**, 437 (1996).
12. C. Locatelli, P. Reschiglian, G. Torsi, F. Fagioli, N. Rossi, and D. Melucci, *Appl. Spectrosc.*, **50**, 1585 (1996).
13. G. Torsi, *Spectrochim. Acta, Part B*, **50B**, 707 (1995).

CHAPTER 4

INSTRUMENTATION

A block diagram for a graphite furnace atomic absorption instrument is given in Figure 4.1. The graphite furnace serves as the atom cell, which converts liquid or solid samples into gaseous atoms. A power supply provides current to control the temperature of a graphite tube between ambient and approximately 3000°C. A light source is used to radiatively excite analyte atoms in the tube. The quantity of light absorbed is recorded by a detection system to do quantitative analysis. The detection system is composed of a wavelength selector, which is used to separate the analytical wavelength from other wavelengths emitted by the source, and a detector, which converts electromagnetic radiation into an electric signal. A signal processor amplifies the signal and sends it to a readout device, which in modern instrumentation is a microcomputer. In addition, the microcomputer serves to control other components of the instrument. A background correction system (not shown in Fig. 4.1) is required to correct for attenuation of the source by molecular absorption or scatter and perform accurate quantitative analysis. Each of the various components is described in detail below, along with a discussion of several commercially available spectrometer designs.

4.1 LIGHT SOURCES

As described in Chapter 2, absorption of light by atoms occurs over an extremely narrow range of wavelengths (0.01–0.05 nm). The emission profile of the source must be narrower than this for the atoms to absorb a significant amount of light, which is required to achieve reasonable sensitivity. The use of a continuum source, such as a deuterium arc or a xenon arc, which emits over wide wavelength ranges (several hundred nanometers), requires a high-resolution monochromator to obtain a sufficiently narrow bandpass. Hence this design, called continuum source atomic absorption, has not yet been incorporated in a commercial instrument. The emission from sources in commercial instrumentation is caused by atoms of analyte, in the source, that are promoted to excited electronic states and subsequently emit atomic lines (Section 2.4). These sources are called line sources. In practice, these lines can

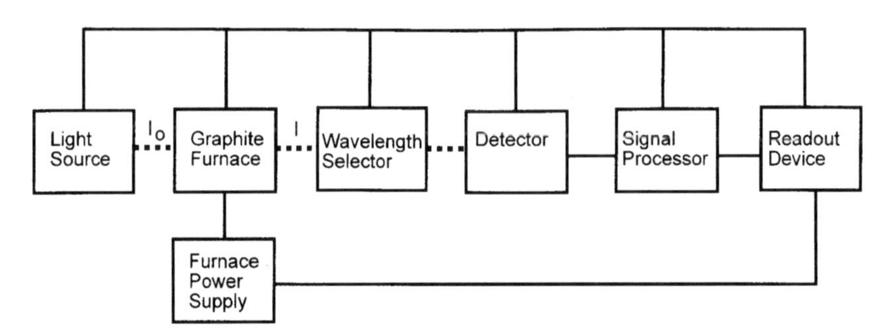

Figure 4.1. Block diagram of instrumentation for graphite furnace atomic absorption spectrometry. The dashed lines represent the optical path and solid lines represent electrical connections.

be considered to be absorbable only by analyte atoms and not by nonanalyte atoms because of their narrow widths. The hollow cathode lamp, first employed by Walsh (1), is the most widely used source for atomic absorption. A second type of line source, the electrodeless discharge lamp, is also commonly used for some elements. Lasers have been employed as light sources for atomic absorption in a few research laboratories. Here one type of laser, semiconductor diode lasers, will be discussed as a light source for GFAAS.

The source is modulated at a specific frequency to discriminate against emission from the graphite furnace or from other chemical species (elements or compounds). The modulation can be performed by either electronic modulation of the source power supply or mechanically by a rotating sector (chopper) to produce an ac system. The amplifier is locked onto the modulation frequency to amplify only signals at that frequency and reject background and noise at other frequencies. The use of an ac system significantly reduces interferences induced by emission.

4.1.1 Hollow Cathode Lamps

The *hollow cathode lamp* (HCL) (2–5), which was first developed by Paschen (6) in 1916 and employed for flame AAS by Walsh (1) in 1955, is the most commonly used source for GFAAS. An HCL is composed of two electrodes, a cathode and an anode, inside a glass envelope (Fig. 4.2). The hollow cathode is composed of a ceramic cylinder coated with the pure metal or a compound containing the analyte. A thick wire, usually composed of tungsten or nickel, serves as the anode. The envelope contains a few torr of a noble gas, usually argon or neon. The application of a potential of 150 to 300 V across the electrodes induces the formation of a plasma within the hollow cathode called a *glow discharge*, with a resulting current of 3 to 30 mA. Free electrons produced in the discharge collide with the fill gas, causing ionization. The noble gas ions

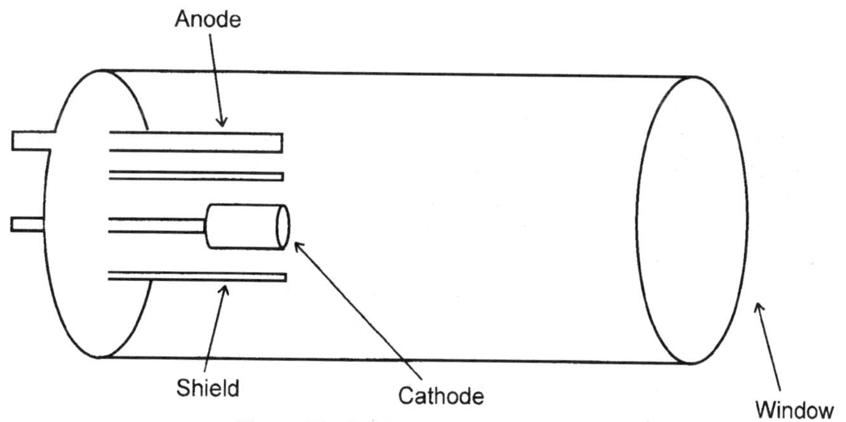

Figure 4.2. Standard hollow cathode lamp.

are electrically attracted to the hollow cathode and the resulting collisions induce vaporization of analyte atoms (*sputtering*). The sputtered atoms subsequently collide with ions and electrons, causing electronic excitation. Lamp emission is caused by the radiative deexcitation of the analyte atoms. The window is composed of silica for elements that emit in the ultraviolet and glass for other elements. An insulating shield is placed between the electrodes to concentrate the emission of light from the hollow cathode region. Hollow cathode lamp emission is at exactly the same wavelength as the absorption line because both involve the same transition. However, because HCLs operate at lower temperatures and pressures than the atom cell (Section 2.6), the emission profile is narrower (typically 0.003 nm) than the absorption profile of analyte in the atom cell (typically 0.01–0.05 nm). The narrow emission profile compared to the absorption profile in the atom cell implies that in most cases nonanalyte atoms cannot absorb HCL light. Other advantages of HCLs include availability for a wide variety of elements, including almost all metals and metalloids, simple operation that involves adjustment of the lamp current to a value prescribed by the manufacturer, high stability, sufficient intensity, and a relatively long lifetime (500–1000 h). The principal disadvantage is that a separate lamp must be purchased for every element determined. Multielement HCLs, which contain more than one element in the hollow cathode, are available, but their intensity is reduced compared to the single-element lamps. Although the absorption signal is not directly proportional to intensity, a reduction in source intensity causes a degradation in the signal-to-noise ratio, and hence in the detection limit and precision. In addition, multielement HCLs may cause a reduction in the linear dynamic range if a nonanalyte line is present within the spectral bandpass of the monochromator (Section 4.3.1). In summary, the use of multielement HCLs is only recommended for rarely determined elements.

As discussed above, relatively high source intensities are desirable for GFAAS because of a reduction in the noise level, which induces an improvement in the limit of detection. The obvious solution would appear to be increasing the current to the lamp. There are two disadvantages associated with this solution. First, a higher operating current reduces the lamp lifetime, and each lamp has a maximum recommended operating current that cannot be exceeded without damage. Second, at high lamp currents, the emission profile of the lamp is altered due to a phenomenon called *self-reversal* (Fig. 4.3). At high HCL currents, the concentration of analyte in the hollow cathode increases, which causes broadening of the source profile. The discharge is expanded, producing ground-state atoms at the edge of the discharge at a relatively low temperature. These atoms may absorb only the center portion of the lamp emission, removing the portion of the lamp discharge that is most effectively absorbed by the atoms in the atom cell. Relatively low levels of self-reversal may decrease the linear portion of the calibration graph (Fig. 4.3*b*); severe self-reversal may prevent measurement of an atomic absorption signal (Fig. 4.3*c*). Self-reversal of a hollow cathode lamp is employed as a method of background correction for atomic absorption (Section 4.6.3).

An additional booster electrode can be used to increase the spectral intensity of an HCL (Fig. 4.4). Atomization of the analyte is performed from the cathode, and excitation occurs at the booster electrode. Separation of atomization and excitation allows increased output from the lamp without self-reversal. A disadvantage of this design is that a separate power supply is required for the booster, which increases the cost.

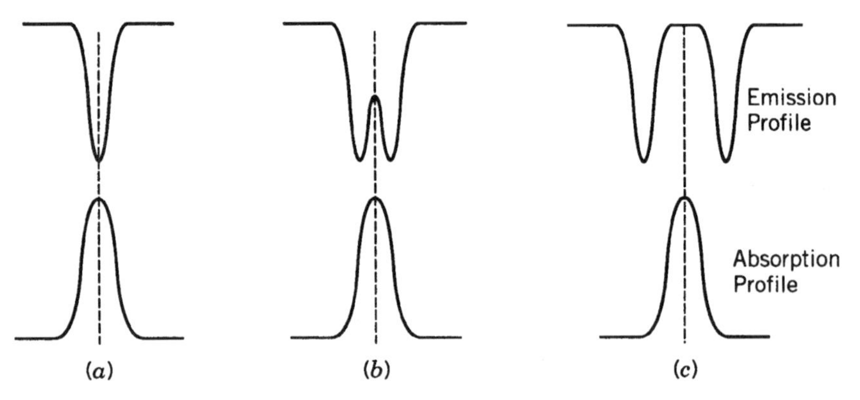

Figure 4.3. Effect of current upon the emission profile of a hollow cathode lamp: (*a*) low current, normal emission; (*b*) medium current, slight degree of self-reversal; and (*c*) high current, complete self-reversal.

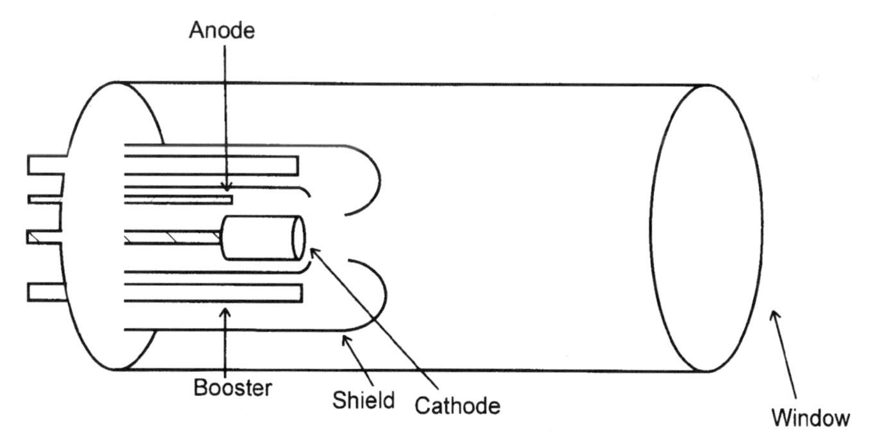

Figure 4.4. High-intensity hollow cathode lamp with booster electrodes to increase excitation efficiency.

4.1.2 Electrodeless Discharge Lamps

Electrodeless discharge lamps (EDLs) (5,7) are alternative line sources to hollow cathode lamps for atomic absorption and have the advantage of higher spectral intensity. They consist of a sample of the analyte, either as the element or a compound (usually a halide), with 0.5 to 5 torr of an inert gas (usually argon) in a quartz sphere surrounded by the coil of a microwave or radio frequency generator, and mounted into an insulated jacket (Fig. 4.5). Microwave (MW) EDLs have the advantage of superior intensity but are unavailable commercially and difficult to operate, and hence have not been commonly employed for AAS. Radio frequency (RF) EDLs are less intense than MWEDLs but are more intense than hollow cathode lamps for several volatile elements, and commercially available, although a separate power supply is required. Particular advantages are obtained for arsenic, selenium, and phosphorous. For these elements, RFEDLs provide an improvement in detection limit compared to HCLs by a factor of 2 to 10. The HCLs of many volatile elements have short lifetimes, while RFEDLs are highly stable with long lifetimes. The principal disadvantage of RFEDLs is the additional expense of a separate power supply ($2000–$3000).

4.1.3 Xenon Arc Lamps

A few graphite furnace AAS instruments have been constructed that employ a high-pressure xenon arc lamp (2,8–11) as the excitation source. Xenon arcs have the advantage of emitting intense continuum emission without significant line

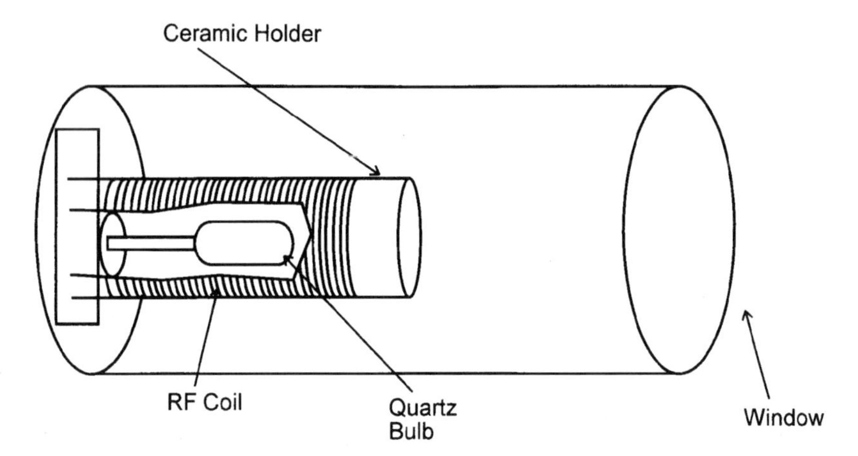

Figure 4.5. Radio frequency electrodeless discharge lamp.

spectra between 200 and 450 nm. Modern lamps are composed of 10 to 30 atm of xenon and two tungsten electrodes inside a silica or sapphire envelope. A parabolic mirror is used to focus the emission into an intense beam. Xenon arc lamps are very intense in the visible region, but their output is considerably lower in the ultraviolet, particularly below 250 nm. The advantage of a continuum source is that combined with a multichannel detection system, such as an échelle monochromator with multiple photomultiplier tubes or a photodiode array, multielemental AAS is possible with a single source. The principal drawback is that a high-resolution monochromator is required to provide a sufficiently narrow linewidth to give comparable sensitivity to line source excitation, and hence continuum source excited GFAAS has been limited to laboratory-constructed instruments.

4.1.4 Semiconductor Diode Lasers

Lasers have been investigated as sources for atomic spectroscopy since the 1960s to develop high-sensitivity methods of analysis. However, the practical application of laser-based analytical atomic spectroscopy has been limited by the high cost (>$100,000), large size difficulty of operation, and poor reliability of the laser systems. *Semiconductor diode lasers* (12–15) offer low cost (<$1000), high reliability, small size ($300 \times 300 \times 150\ \mu m$), and easy operation, making them potentially useful sources for GFAAS (12). They can also be tuned on and off an atomic line (0.002–0.02 nm) by variation of the current applied to the diode at frequencies up to several gigahertz. Each type of laser diode covers a wavelength range of 10 to 40 nm.

At the present time, the biggest limitation of laser diodes is their limited wavelength range. Commercial laser diodes are available from 630 to 1600 nm, and by the use of second-harmonic generation techniques, wavelengths from 315 to 500 nm. The limited wavelength range means that some elements cannot be determined with these sources and that others must be determined using less sensitive wavelengths. However, these devices have considerable potential for high-sensitivity analysis because of their unique combination of high intensity, narrow linewidth (5×10^{-5} nm), and rapid tunability provides particular advantages for the use of *wavelength modulation.*

Wavelength modulation (WM) has been employed to correct for nonanalyte absorption induced by molecules or scatter of the source beam (background correction—Section 4.6) (16,17). In WM, the laser is first tuned on the atomic absorption wavelength, where analyte and background absorption may occur, and then off the wavelength by 2 to 20 pm (depending on the width of the atomic line) where only background absorption may occur. Subtraction of the on-wavelength—the off-wavelength measurement gives a background-corrected signal. Wavelength modulation has been shown to provide a significant improvement in the detection limit for laser atomic absorption because the signal-to-noise ratio is improved by the use of an intense, narrow-band source, allowing measurement of absorbances as low at 10^{-4} to 10^{-6}. Laser diodes offer considerable potential for GFAAS, although commercial development of these sources for GFAAS will probably not occur until the entire ultraviolet-to-visible range is accessible by these sources.

4.2 GRAPHITE FURNACE

Since the initial use of a graphite furnace (18–25) with atomic absorption spectrometry by L'vov in 1959 (18–20), considerable development in graphite furnace materials and design, called modern furnace technology, has occurred that has increased the sensitivity and accuracy of the atomizer for practical analysis. This section focuses on a description of successful furnace design. Readers interested in a more historical perspective are referred to Appendix A and other sources (4,5,21–27).

The graphite furnace serves as the atom cell, whose function is to convert the analyte in a sample (solid, liquid, or gas) into gaseous atoms that can be monitored spectroscopically. A schematic diagram of this device, which is also called an electrothermal atomizer (ETA), is given in Figure 4.6. A graphite tube, typically 3 to 6 mm in inner diameter and 20 to 40 mm in length, is heated resistively by a high current (up to several hundred amps), low voltage (6–12 V) ac power supply. Sample introduction into the tube is performed through the dosing hole in the center of the furnace length. The source beam passes through the tube in the axial direction and on to the detection system. The temperature of the tube can be controlled from ambient to 2700°C with a precision of ±10°C up to 200°C and ±50°C at the higher end of the range. The tube is mounted in graphite inserts that are held in place by brass or stainless steel water-cooled electrodes that are electrically connected to the power supply. The internal and external walls of the tube are bathed in a purge gas, usually argon, to exclude oxygen and prevent combustion of the graphite surfaces. Argon is preferable to nitrogen because the latter forms compounds with several elements (e.g., aluminum) and toxic CN gas. Removable quartz windows serve to prevent intrusion of oxygen into the furnace through the optical path. The internal flow of gas is usually turned off (gas-stop) when the analytical measurement is performed to maximize the time that the analyte atoms are present in the tube (residence time) and the sensitivity. The internal gas flow generally may be specified by the analyst during each step of the furnace heating cycle (Section 4.2.2). In addition, it is usually possible to switch to an alternative gas (e.g., air or oxygen for oxygen ashing, Section 5.1.2) at specified times.

The temperature of the tube is controlled two ways. During the majority of the furnace cycle, the tube temperature is regulated by the amount of voltage (current) applied. However, in order to obtain the best sensitivity and accuracy, the furnace is usually heated with the maximum possible heating rate (>1000°C/s) to a temperature just above the minimum required to completely atomize the analyte. An *optical temperature sensor* is used to monitor the tube wall temperature in most modern commercial designs during maximum power heating. When the tube temperature exceeds the selected atomization temperature, the optical pyrometer turns off the maximum power heating, and temperature control is returned to the furnace power supply. Alternatively, some instruments use the optical temperature sensor to control the temperature of the tube during the pyrolysis and atomization steps (Section 4.2.2).

4.2.1 Graphite Tube Material and Design

Early GFAAS work in the 1970s employed relatively large, longitudinally heated tubes (50 mm in length, 8 mm in diameter) that had the advantages of being able to accommodate large sample volumes and relatively easy alignment of the source through the tube. Longitudinally heated furnaces are commonly referred to as Massmann furnaces, who first described this design in the late 1960s (28). However, large furnaces cannot be heated rapidly and tend to heat unevenly, which allows condensation of material in the cool areas. During the early period, open graphite atomizers (Fig. 4.7) were also employed for AAS. These atom cells provide very high sensitivity for simple aqueous solutions but are unsuitable for analysis of complex sample matrices. Atomization occurs into

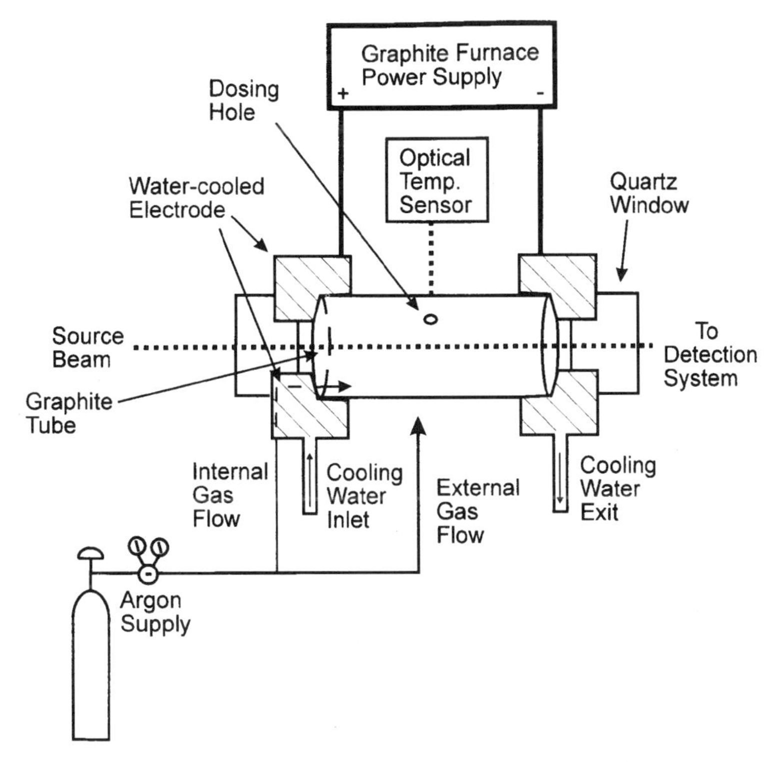

Figure 4.6. Schematic diagram of a graphite furnace atomizer.

a relatively cool environment that induces molecule formation between the analyte and its matrix.

The most commonly used material for early graphite tubes was *polycrystalline graphite* (electrographite), which is porous, allowing diffusion of analyte molecules into the material, and is relatively reactive toward metals, which may cause interferences. The porosity and reactivity of electrographite may induce tailing, memory effects, or incomplete atomization. This material is particularly unsuitable for the determination of several elements (e.g., vanadium, titanium, molybdenum) that form stable, involatile carbides.

Modern graphite furnace systems employ graphite tubes (20–30 mm in length; 3–6 mm in diameter) that can be rapidly heated (>1000°C/s) to reduce interferences. A high heating rate also allows vaporization with a lower final atomization temperature, which is particularly advantageous for involatile elements. The problems with electrographite can be greatly reduced by the formation of a layer of pyrolytic graphite (50 μm) on the surface of a polycrystalline tube. *Pyrolytically coated graphite* tubes are produced by heating

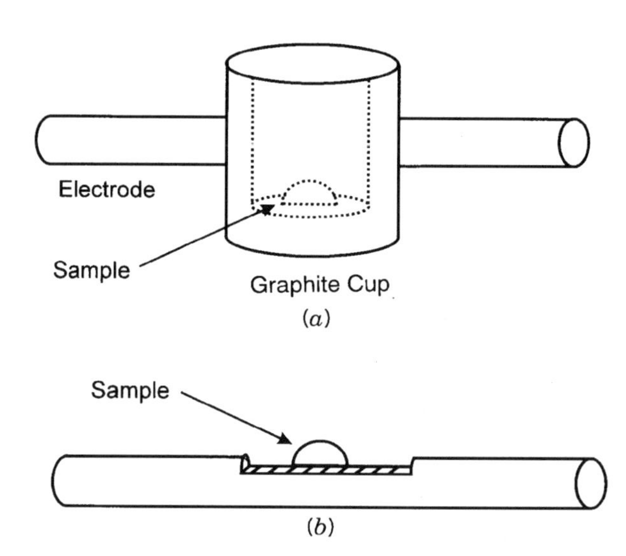

Figure 4.7. Open atomizers for atomic absorption: (*a*) graphite cup and (*b*) graphite rod.

an electrographite tube in 5% methane/argon. The dense layer of pyrolytic graphite serves to reduce diffusion of analyte into the graphite and chemical reactivity. Carbide-forming elements should only be determined with pyrolytically coated tubes to obtain acceptable sensitivity and accuracy.

A variety of other materials have been employed as tube substrates for graphite furnace AAS, which include total pyrolytic carbon, glassy carbon, metal furnaces (e.g., tantalum, tungsten), metal liners (thin metal sheets), and metal impregnated into graphite (5,29–32). In general, these approaches have not been proven to be superior to pyrolytically coated tubes for a wide variety of applications, and hence they have been employed in a relatively small number of laboratories.

Up until recently, all commercial graphite furnaces were heated by a longitudinal flow of current through the furnace. Welz and co-workers (33) showed that the temperature of the gas phase in a longitudinally heated graphite furnace may vary as much as 1200°C from the center to the ends. The relatively cool temperatures at the ends of the furnace, caused by contact with the water-cooled electrodes, may allow analyte atoms to condense or analyte-containing molecules to form in the colder areas. Condensation has been shown to induce chemical interferences (Section 5.2).

One approach to minimize these effects is the use of a transversely heated graphite atomizer (THGA) with integrated contacts (34) (Fig. 4.8). The temperature gradients are reduced with these atomizers compared to longitudinally heated furnaces, resulting in reduced interferences compared to the

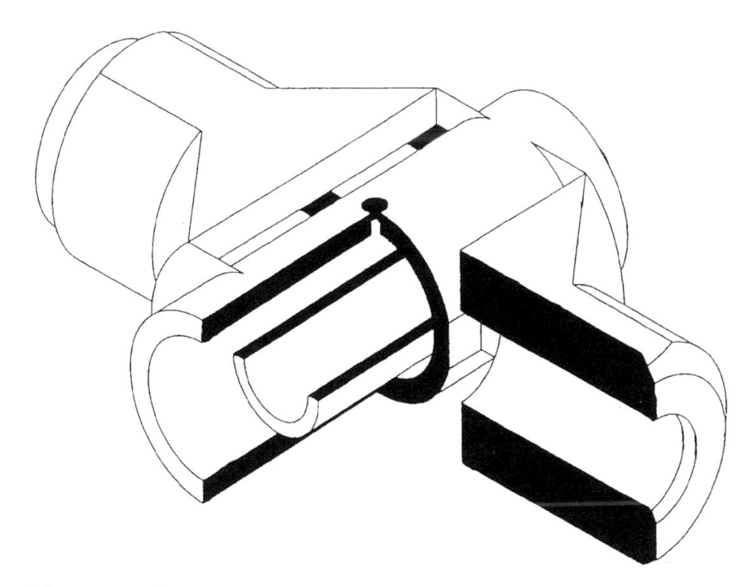

Figure 4.8. Transversely heated graphite tube with integrated platform and contacts. Taken with permission from Reference 4.

Massmann design (35). However, Frech and co-workers (36,37) showed that analyte and matrix may condense in cool regions in a THGA. In addition, commercially available THGAs (Perkin-Elmer Corporation, Norwalk, CT) are employed in a longitudinal magnetic field for ac Zeeman effect background correction (Section 4.6.4.1) that has dictated the use of a relatively short tube length (18 mm) that resulted in a reduction of the peak area sensitivity by a factor of 2.

In order to reduce interferences and improve the sensitivity, pyrolytically coated carbon disks with small (3.2 mm) apertures, called end caps, have been inserted in the end of the tube (37,38). The use of end caps reduces the rate of diffusion of the analyte out of the tube and provides comparable peak area sensitivity to longitudinally heated furnaces. In addition, they increase the temperature of the gas phase at the ends of the tube, reducing condensation and the potential for interferences.

4.2.2 Furnace Heating Cycle

In order to make an analytical measurement by GFAAS, it is necessary to set up a program to control the temperature of the furnace following introduction of the sample. An optimized temperature program allows vaporization of nonanalyte (matrix) species, such as solvent and organic matter, before atomization of the analyte, in order to reduce potential interferences. It is also essential that all analyte vaporize in a temporally reproducible way, and that all is removed by the end of the furnace program.

A typical furnace program consists of the following steps: (1) *sample introduction*, (2) a *dry step*, to remove solvent (usually water) from the sample, (3) a *pyrolysis step* (also called char or ash) to remove organic and other volatile materials in the sample before the analyte is vaporized, (4) a *cool-down step* to allow the furnace to reach ambient temperature, (5) an *atomization step* in which the analyte is atomized and the integrated absorbance recorded, and (6) a *clean step* to remove any residual material from the graphite tube. A schematic diagram of a furnace program is shown in Figure 4.9, and typical temperatures and times employed are listed in Table 4.1.

With modern instrumentation, sample introduction is normally done with an autosampler. For the introduction of liquid samples, which are used in the vast majority of GFAAS work, the autosampler consists of a mechanical arm attached to a piece of plastic tubing and a series of standard and sample solutions in cups positioned in a sample tray. One end of the tubing is connected to a pump and a deionized water supply, while the other end is connected to a mechanical arm that may be inserted into the solutions. The sample tray rotates and moves laterally to sample each solution in the tray. The tubing is inserted into a sample, the pump draws a user-specified volume into the tubing, and the

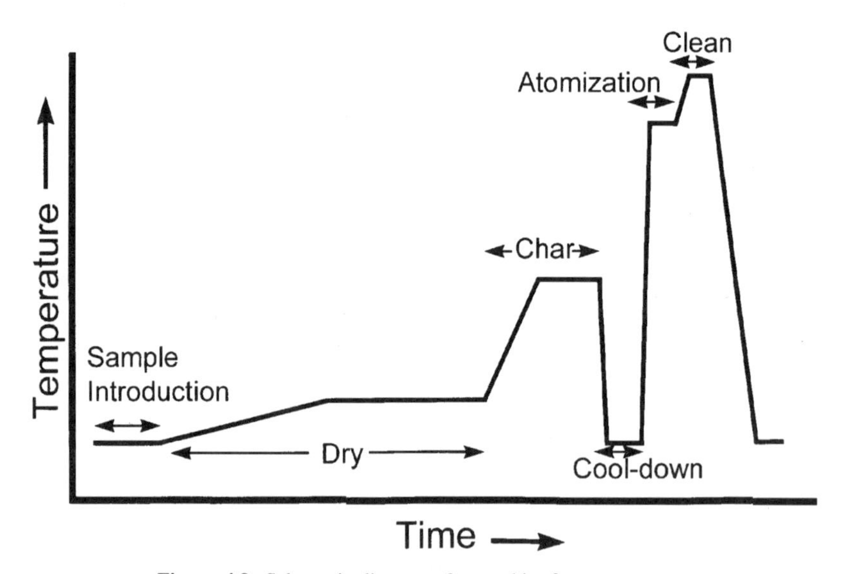

Figure 4.9. Schematic diagram of a graphite furnace program.

Table 4.1 Typical Graphite Furnace Program Temperatures and Times

Cycle Step	Tube Temperature, °C	Temperature Ramp, °C/s	Hold Time, s
Dry	100–150	2–20	20–40
Pyrolysis	300–1500	50–200	10–30
Cool Down	200	−100–500	5–15
Atomize	1600–2700	Maximum power heating	5–10
Clean	2500–2700	200–500	3–10

mechanical arm moves the tubing through the dosing hole into the graphite tube to deliver the sample. Generally sample volumes of 5 to 50 µl are employed; most modern programs use 10 to 25 µL. Most autosamplers allow introduction of up to three solutions into the furnace [e.g., sample, standard, chemical modifier (Section 5.2)]. The autosampler then rinses the tubing with deionized water to clean it, and the heating cycle of the furnace begins.

It is usually necessary to observe the sample introduction process the first few samples of the day to ensure the solution is precisely and accurately delivered into the furnace. The end of the tubing normally should pass through the center of the dosing hole and be positioned 2 to 3 mm above the graphite substrate during sample introduction. Normally the position is set manually by the user, before a graphite furnace cycle is initiated. A dentist's mirror, or a solid-state charge-coupled device (CCD) camera available on some instruments, is used to observe the sample introduction process (39,40). The CCD camera has been used to optimize set-up and alignment of the graphite tube and to ascertain that all of the liquid has been deposited correctly in the tube. Phase transitions of the injected samples may be observed during the dry and pyrolysis steps.

A second type of autosampler employs deposition of the sample as an aerosol spray (41). This autosampler, marketed by Thermo Jarrell Ash (Franklin, MA) as the "Fastac," has the advantage that it can also be used for flame AAS and is well-suited for use with a graphite tube that has been preheated, which decreases the time required for the dry step and hence the analysis time. However, aerosol deposition requires several milliliters of sample (although only 20–50 µl are introduced into the furnace), and is problematic with samples that are viscous or contain undissolved material.

The use of an autosampler has three principal advantages for GFAAS. Precision is usually degraded with manual pipetting because it is difficult to reproducibly dispense the sample in the same location. Manual pipetting may cause introduction of material on the edge of the dosing hole of the tube. Typical precision values of several percent are obtained with manual sample introduction, as compared to 1% or less with an autosampler. Second, each heating cycle requires 2 to 3 min, and hence manual operation is unproductive for the operator. Third, pipette tips employed in manual pipetting may introduce contamination (42).

The dry step serves to remove liquid present in the sample. With aqueous samples, the furnace is heated to 110 to 150°C; lower temperatures are used with wall operation and higher temperatures with a probe or platform (Section 4.2.3). The furnace is heated relatively slowly (2–20°C/s) to a temperature just high enough to completely remove the solvent without spattering. Improper drying can lead to poor precision. It is normally prudent to observe the drying step for the first few furnace cycles of the day, using the mirror or camera system described above.

The pyrolysis step serves to remove nonanalyte (matrix) components of the sample. A moderate heating rate of 50 to 200°C/s is generally used during this process. It is usually desirable to use as high a pyrolysis temperature as possible without vaporization of the analyte, and determination of the optimum temperature is generally required for each element and type of sample. Pyrolysis temperature optimization involves measuring the absorbance from a standard or sample at a fixed atomization temperature with a variety of pyrolysis temperatures. Usually it is necessary to do temperature optimization for standards and samples to ensure the pyrolysis temperature is optimum for both. Figure 4.10*a* shows an example of a pyrolysis temperature optimization with optimum at 1100 to 1200°C. It is usually desirable to use a pyrolysis temperature of at least 1000°C to minimize interferences. The use of chemical modifiers (Section 5.2) allows the use of relatively high pyrolysis temperatures, even for volatile elements (e.g., lead).

The introduction of air or oxygen into the furnace during this step allows ashing in the graphite tube (oxygen ashing). Oxygen ashing allows more complete vaporization of some organic matrices, such as blood, and prevents formation of carbonaceous residue that cause interferences (Section 5.1) (43,44). Oxygen ashing is performed at temperatures below 800°C to prevent combustion of the furnace.

Following the pyrolysis step, the furnace temperature is returned to ambient using a cool-down step. The use of a cool-down step helps ensure atomization of the sample into a hot environment, which has been shown to reduce interferences (Section 4.2.3).

In the atomization step, the temperature of the graphite tube is rapidly increased with maximum power heating (>1000°C/s) to just above the temperature required to atomize the analyte (1600–2700°C). The analytical measurement is performed during the atomization step, and as described in Sections 2.8 and 2.9 and illustrated in Figure 2.8, a transient absorption signal is obtained. Optimization of the atomization temperature is generally required for

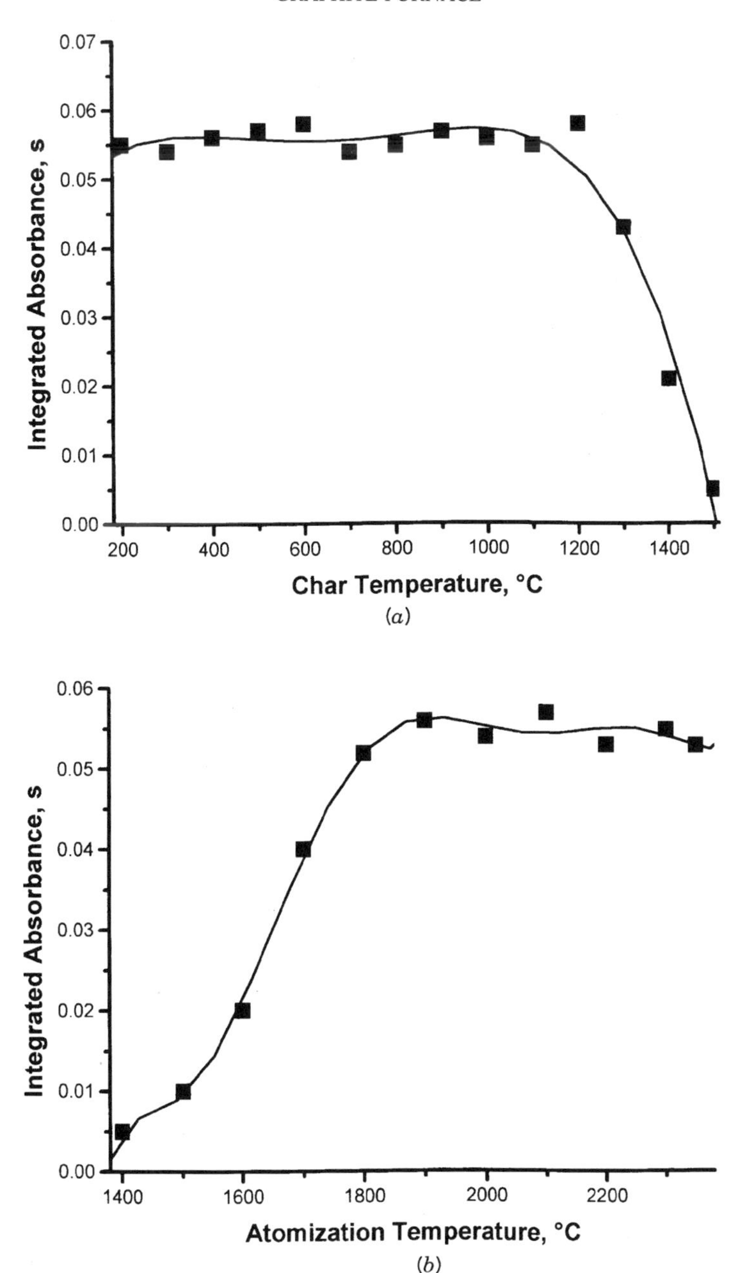

Figure 4.10. (*a*) Pyrolysis temperature optimization graph. The optimum pyrolysis temperature is at approximately 1100°C. (*b*) Atomization temperature optimization graph. The optimum atomization temperature is at approximately 1900°C.

standard solutions and samples and involves the measurement of absorbance of a standard or sample with a fixed pyrolysis temperature and a series of atomization temperatures. Figure 4.10*b* shows an example for which the optimum temperature is 1900°C.

The clean step serves to remove residual material from the sample in the furnace. The tube is usually heated at a few hundred degrees per second up to 2500 to 3000°C. After the clean step, the tube is allowed to cool to ambient temperature, and the cycle is initiated again.

Typical furnace programs require 2 to 3 min per analysis. *Fast furnace programs* have been developed in order to approximately double the sample throughput of GFAAS (45–49). Halls (49) has reviewed the use of fast furnace programs for GFAAS. Typically the sample is introduced into a hot (150–200°C) furnace, and short (5-s) pyrolysis steps are used at relatively high temperatures (>1000°C). Good results have been obtained for a number of elements in relatively easy matrices. More work needs to be performed to determine whether fast programs can be widely used for routine analysis.

Instrument manufacturers provide several furnace programs for each element that may be used as starting points for analysis (Section 7.4, Appendix C). The analytical literature (Appendix B) may also be consulted for initial conditions.

4.2.3 Methods of Atomization

Early commercial atomic absorption atomizers heated at relatively slow rates (500–800°C/s) with sample introduced on the wall of the tube, and hence atomization occurred as the furnace was heating to its final temperature. Under these conditions, atomization occurred into a tube whose temperature varied from furnace cycle to furnace cycle and along the length of the tube, with the center several hundred degrees hotter than the ends (33). Wall atomization was shown to be less suitable for real sample analysis with volatile elements because these temperature variations were shown to degrade precision and induce the formation of analyte-containing molecules that cause chemical interferences (Section 5.2), although wall atomization is preferable for extremely involatile elements. Several approaches have been investigated to ensure atomization occurs into a relatively high temperature environment.

The most commonly used approach is the *L'vov platform* (50) in which a sample is introduced onto a small graphite shelf (usually pyrolytically coated polycrystalline graphite or totally pyrolytic graphite) inside the tube (Fig. 4.11). Current does not pass directly through the platform, and hence it is primarily heated radiatively by the tube walls. Consequently, the use of a L'vov platform with a rapidly heated furnace ensures that atomization occurs after the tube and

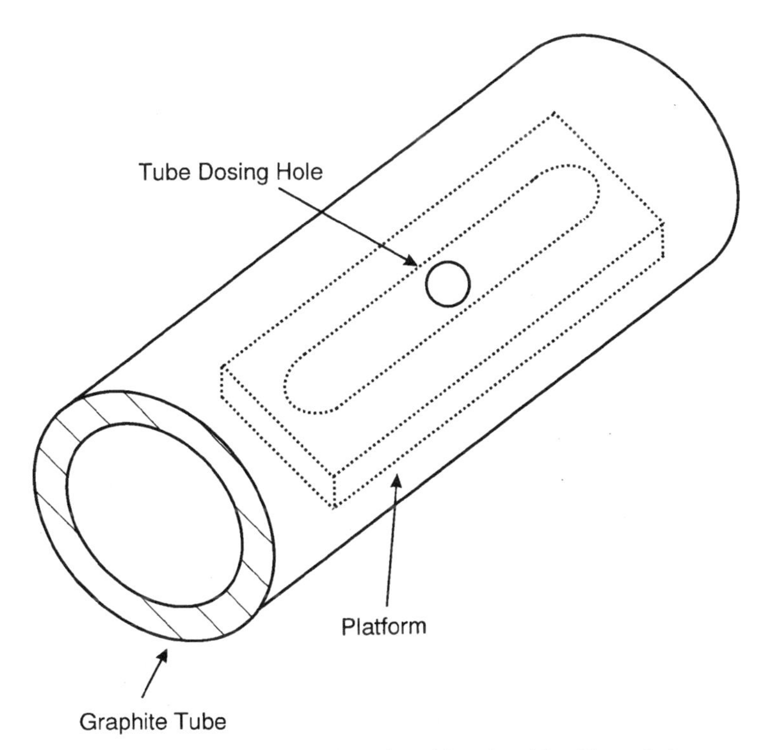

Figure 4.11. Longitudinally heated graphite tube with a L'vov platform.

the gas inside it have reached a relatively constant temperature after maximum power heating. This means that atomization occurs into a hot environment, minimizing interferences. Today most manufacturers offer graphite tubes with integrated platforms. Transversely heated graphite tubes include a pyrolytically coated platform machined from the same piece of graphite (Fig. 4.8). Contact between the tube and platform is minimized to reduce heat transfer through direct contact.

A second approach is the *delayed atomization cuvette* in which a graphite tube is modified so that the outer diameter at the middle is greater than at the ends, with a constant inner diameter (Fig. 4.12) (51,52). In a delayed atomization cuvette, sample is introduced into the middle, thicker region. The thinner ends of the furnace are heated more rapidly than the center, allowing vaporization into a relatively hot environment. In general this approach does not seem to be as effective at reducing interferences as the L'vov platform.

Probe atomization involves the use of a graphite probe that is inserted into and removed from the tube by a stepper motor (Fig. 4.13) (53). Sample is deposited onto the probe with the probe inside the furnace, and the sample is

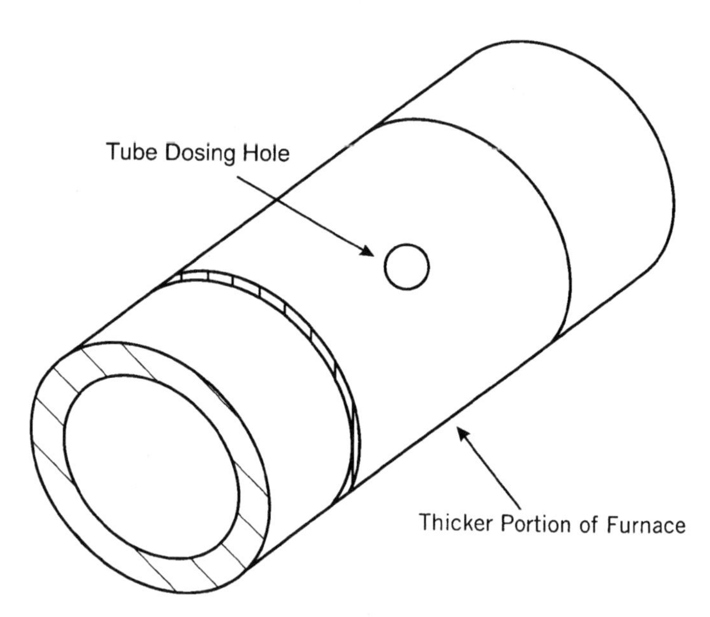

Figure 4.12. Delayed atomization cuvette.

dried and ashed. The probe is then withdrawn from the furnace, which is subsequently heated to the atomization temperature. The probe is rapidly reinserted into the furnace, allowing atomization into a hot environment. Probe atomization has not been as widely employed as platform atomization, probably because of the added complexity of the instrumentation, and because the insertion of the cool probe into a hot tube cools the vapor, which prevents isothermal vaporization. In addition, the probe hole provides an additional avenue for loss of analyte.

A *two-step furnace* employs two power supplies, one to heat the graphite tube transversely, and the other to heat a graphite cup, just below an aperture in the tube, into which sample is introduced (54,55) (Fig. 4.14). The tube is heated to the atomization temperature, and subsequently the cup is heated to vaporize the analyte into the isothermal tube. Interestingly, this design is very similar to the first graphite furnace instrument described by L'vov (18–20). The design has not been available in commercial instrumentation, probably because of the additional cost of two power supplies, and has relatively small advantages for most analytical applications compared to conventional atomization with a transversely heated furnace, although two-step furnaces have been employed for fundamental studies.

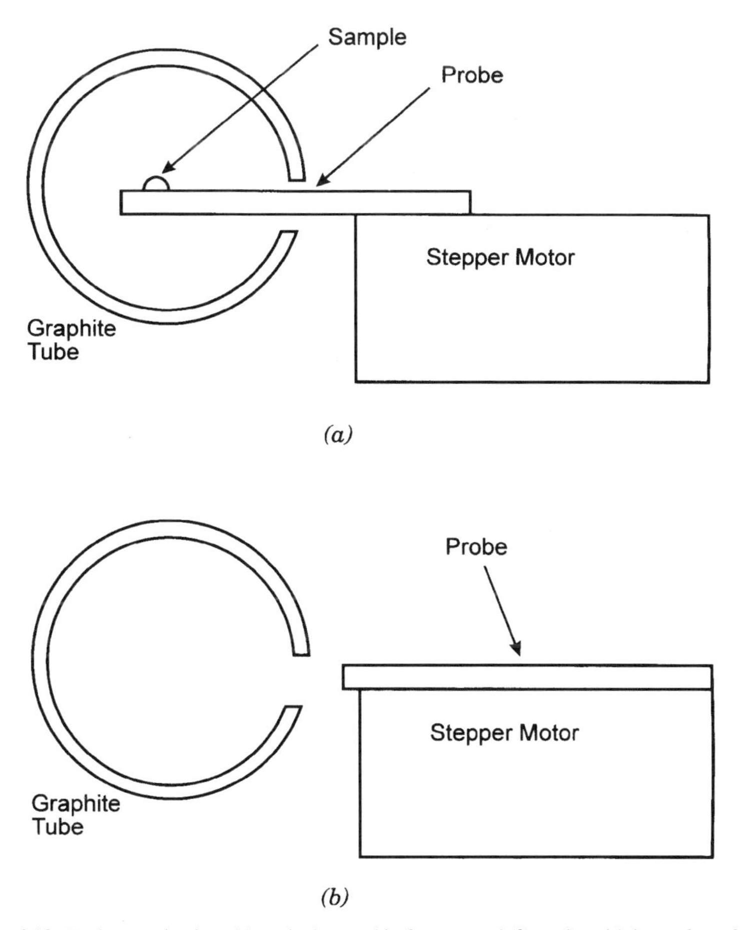

Figure 4.13. Probe atomization: (*a*) probe inserted in furnace and (*b*) probe withdrawn from furnace.

4.3 WAVELENGTH SELECTORS

With line source excitation, the function of the wavelength selector in atomic absorption is to separate the analytical line from other lines emitted by the light source (Fig. 4.15). Two basic types of wavelength selectors are commonly used in analytical instrumentation: monochromators and filters. Virtually all commercial GFAAS instruments have been equipped with relatively low resolution monochromators. An échelle monochromator has recently been incorporated in a multielemental instrument.

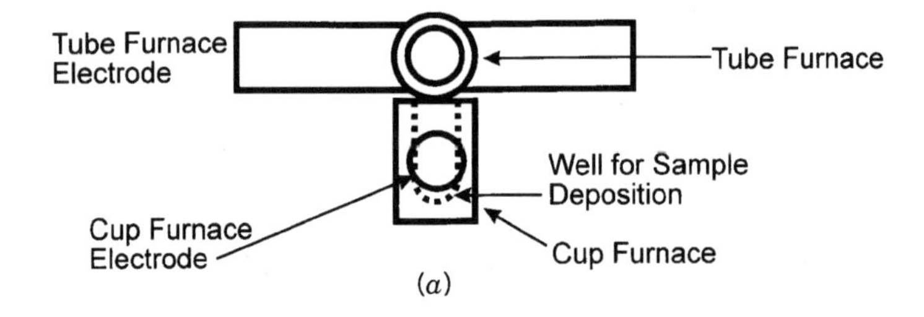

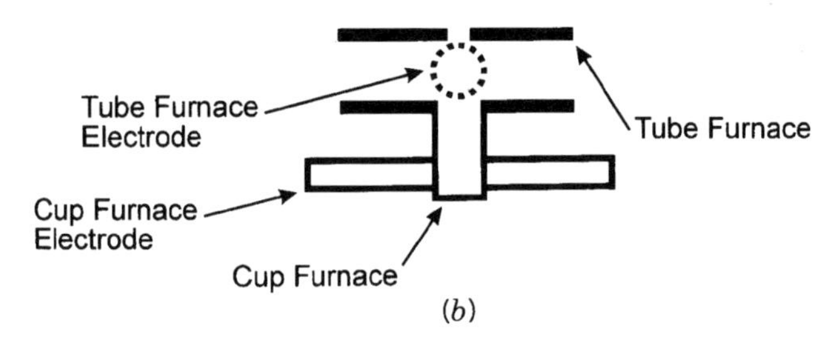

Figure 4.14. Two-step furnace: (*a*) front view and (*b*) side view.

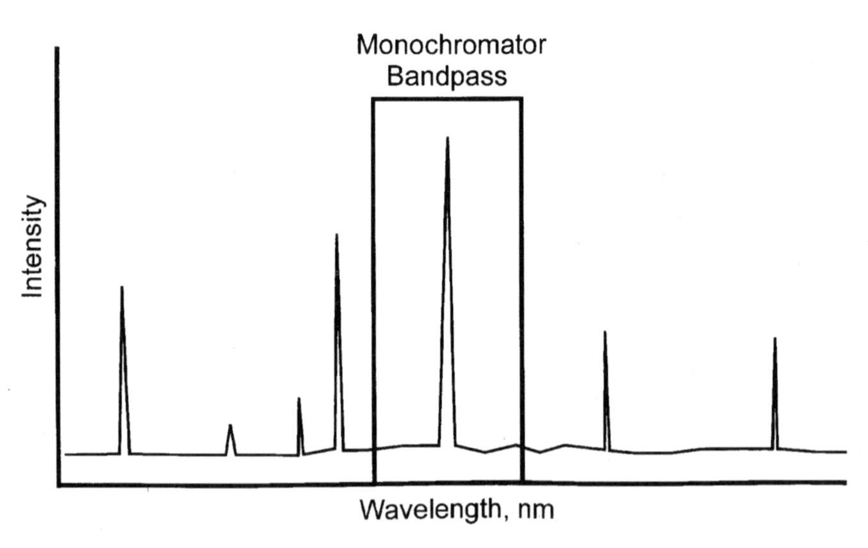

Figure 4.15. Role of the monochromator in AAS: isolation of the resonance HCL emission line.

4.3.1 Monochromators

With a line source, the monochromator must be able to separate the analytical line from any other lines emitted by the source. A diagram of a typical one-dimensional monochromator is shown in Figure 4.16. Light enters the monochromator through an entrance slit, is converted to a parallel beam of radiation by a collimator, is dispersed into various wavelengths by an échellette grating, and is focused onto the exit slit by a second collimator.

The resolution, or ability of a monochromator to separate spectral lines, may be specified in terms of the *spectral bandpass*, which is defined as one-half of the wavelength distribution passed by the exit slit. In general, in order to completely resolve two atomic lines, the spectral bandpass must be no greater than one half their difference in wavelength. In other words, if two lines are separated by 2 nm, than the spectral bandpass must be 1 nm or less. Further details on the operation of monochromators is given in analytical textbooks (17,56–58).

The bandpass required for AAS is dependent on the emission profile of the HCL for the analyte. Some elements, such as copper, calcium, and magnesium, have relatively few emission lines, and none close to the resonance wavelength, allowing the use of a relatively wide spectral bandpass of 1 nm. Other elements, such as nickel and iron, have many lines and require narrower bandpasses. In general, a bandpass of 0.2 nm is sufficient. The use of a narrower spectral bandpass than required to isolate the emission line will result in a degradation of the signal-to-noise ratio and the precision. The use of an excessively wide bandwidth will cause a reduction in the sensitivity (slope of the calibration

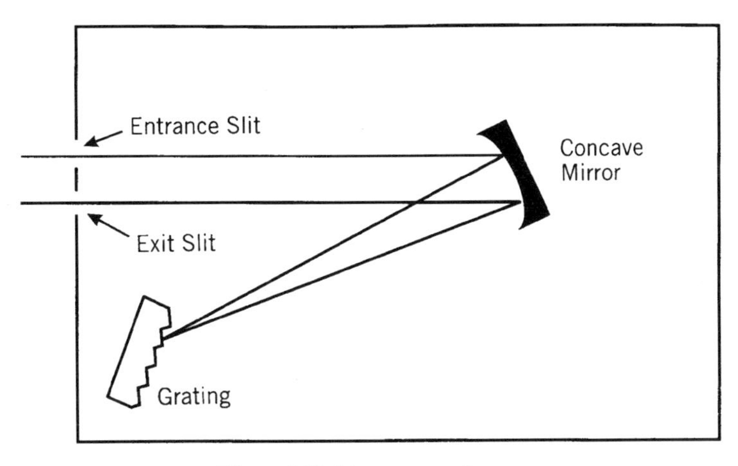

Figure 4.16. Littrow monochromator.

curve) and reduce the linear range of the calibration curve. This phenomenon occurs because the second line reaching the detection system does not absorb light, and the maximum amount of light absorbed is equal to the intensity of the unabsorbable line divided by the intensities of the two lines.

In practice, atomic absorption manufacturers provide recommended spectral bandwidths for each element and analytical wavelength in their method manuals (Appendix C and AAS cookbooks). These values should provide good performance for most routine analysis with single-element hollow cathode lamps. However, it may be necessary to use a smaller bandpass with multielement HCLs. If the bandpass is excessively wide, excessive curvature in the calibration graph will be present that may be significantly improved by the use of a narrower bandpass.

4.3.2 Échelle Monochromators

A multielement GFAAS instrument employs an *échelle monochromator* to allow simultaneous analysis of several elements with high throughput per unit spectral bandpass in a relatively compact design (17,56,58,59). This design, which has been commonly used in atomic emission instruments, employs a large angle of diffraction and high orders in order to improve resolution. However, it is necessary to use a large number of orders (e.g., 90) in order to cover the UV–visible region from 200 to 800 nm. A second dispersing element (usually a prism) is required to separate overlapping orders of the grating in a second dimension (Fig. 4.17*a*). The result of this arrangement is a two-dimensional spectrum at the focal plane (Fig. 4.17*b*). The resolution is approximately linear at any given order but is worse at lower orders or higher wavelengths. This is not a significant disadvantage for atomic absorption because the majority of spectral lines are in the UV.

4.4 DETECTORS

The detector serves to convert radiation transmitted through the atom cell into an electrical signal. The vast majority of GFAAS instrumentation has employed *photomultiplier tubes* (PMTs) as detectors because of their high sensitivity, reliability, and ease of operation. A PMT includes a cathode composed of material that releases electrons when exposed to light (a photoemissive material) and a wire anode inside a quartz envelope maintained in a vacuum (56,58). The cathode is usually maintained at 500 to 1500 V negative compared to the anode. In addition, there are 5 to 15 additional electrodes called dynodes, which are maintained at potentials intermediate to the cathode and anode. Light from the monochromator enters the window and strikes the photoemissive cathode,

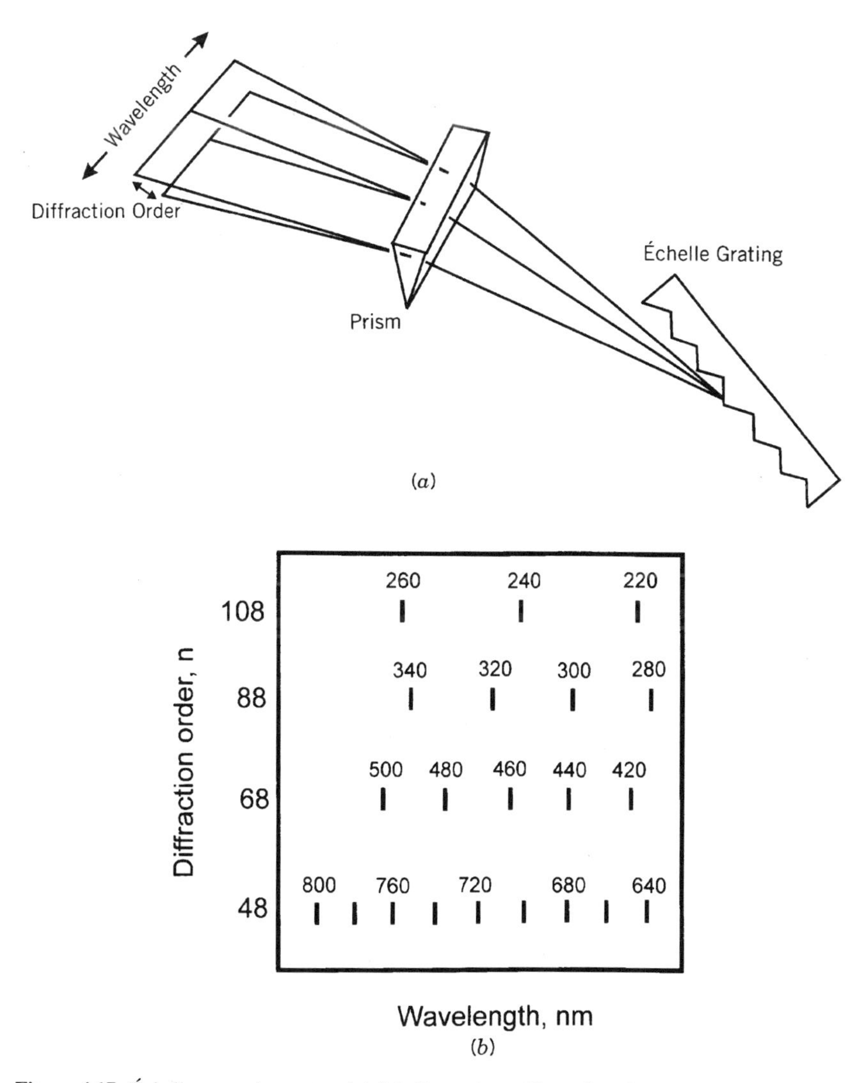

Figure 4.17. Échelle monochromator: (*a*) échelle grating with a prism for cross dispersion and (*b*) schematic diagram of the focal plane of an échelle monochromator showing the location of wavelengths for 5 of 70 orders.

releasing several (3–10) electrons per photon. These electrons are then electrically attracted to the first dynode, which is maintained at a more positive potential than the cathode. The resulting collision causes the ejection of several more (3–10) electrons per incident electron. These electrons are then attracted to the second dynode, causing similar gain, and so forth until the electrons reach the anode. Hence, for a PMT with nine dynodes, a gain of 10^6 to 10^7 is achieved for each incident photon. In general, an ultraviolet sensitive PMT will provide high sensitivity for wavelengths between 190 and 600 nm, which covers the majority of elements. A few elements, such as potassium and cesium, are commonly determined at longer wavelengths and require a red-sensitive PMT for best sensitivity. The relatively large size and high cost of PMTs has limited the use of multiple PMTs for multichannel analysis, although a few instruments have been constructed with more than one PMT.

Multichannel detectors consist of an array of photodetectors arranged so that the entire range of wavelengths dispersed by a grating can be detected simultaneously. Traditionally, two major types of multichannel detectors have been used in analytical instrumentation, photodiode arrays and image-sensing vacuum tubes (vidicons), but these devices cannot match PMTs in terms of sensitivity and consequently have not been commonly used as detectors for AAS. Since the mid-1980s, a considerable amount of work has focused upon the use of a new type of multichannel detector, called charge-transfer devices, whose performance characteristics are comparable to those of PMTs (60–62). Each charge transfer device consists of a two-dimensional array of metal-oxide-semiconductor (MOS) capacitors manufactured onto a silicon chip. In one commercial design, the output of an échelle monochromator is imaged onto a charge-transfer device to allow simultaneous determination of up to six elements.

4.5 SIGNAL PROCESSING, READOUT, AND INSTRUMENT CONTROL

The electrical signal from the detector is converted into absorbance by use of a logarithmic amplifier. It is important to accurately sample the transient atomic absorption signal produced in a furnace with an adequate sampling frequency (generally 100 Hz or more). McNally and Holcombe (63) proposed that higher frequencies (1000 Hz or more) may be required to accurately represent the signals of elements that are rapidly produced during the atomization step. A sufficiently fast time constant (1–10 ms) must also be employed to prevent excessive smoothing of the transient signal. Considerable distortion of GFAAS signals was observed with some early spectrometers in the 1970s because electronics were employed with slow response times.

Readout from modern GFAAS instrumentation is performed by a built-in or external microprocessor. Most instruments provide both integrated and peak absorbance values, and highly resolved digital display of atomic absorption signals. The video display may include uncorrected signals, background, and background-corrected signals (Section 4.6), as well as the temperature of the graphite tube (Fig. 4.18).

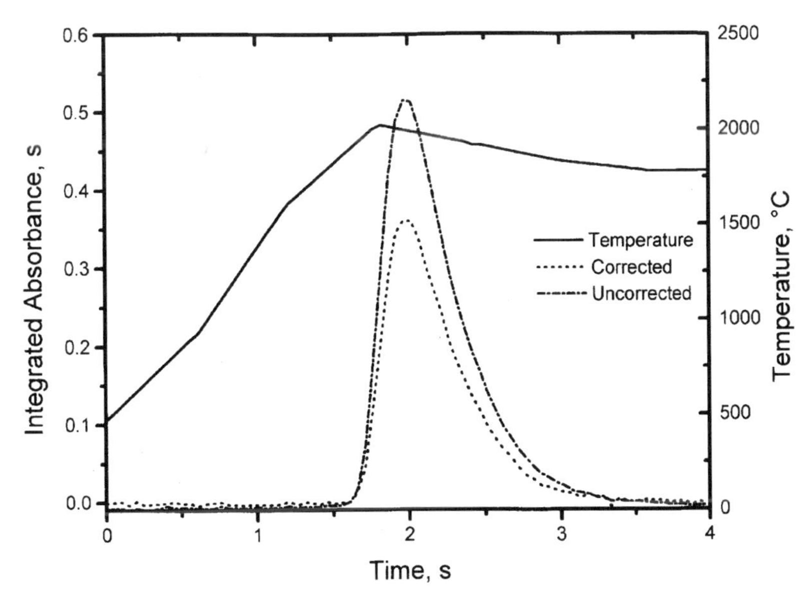

Figure 4.18. Uncorrected and corrected absorbance profiles and temperature from a modern GFAAS instrument for aqueous lead (480 pg) with self-reversal background correction and platform atomization. The atomization temperature was 1800°C, the pyrolysis temperature was 1100°C, and palladium nitrate (2 μg as Pd) and magnesium nitrate (4 μg as Mg) were the chemical modifiers.

In addition to data collection, the microprocessor also serves to control various components of the instrument, including the graphite furnace, autosampler, source, monochromator, and detection system. Instrumental parameters for each of these components are entered via the keyboard and can be stored and recalled for later use. The instruments can collect data from standards in order to produce a calibration graph or can be programmed to do analysis by the method of standard additions (Section 3.3) or recovery checks (Section 7.3). After sample data is collected, the concentration of the analyte may be reported with statistical analysis. Instruments can be programmed to perform quality control checks on a standard or standard reference material with a defined level of accuracy. If the quality control criteria are met, the analysis continues; if not, the instrument may be recalibrated or other steps taken that are defined by the operator. Some instruments can record the total number of furnace cycles on a graphite tube, or the number of hours of use on a hollow cathode lamp. In summary, modern GFAAS instruments can be operated in an automated mode that provides higher accuracy, precision, and speed than manual methods.

4.6 BACKGROUND CORRECTION

Background attenuation, due to molecular absorption or scatter, is a significant interference for the accurate determination of elements in sample matrices. Several methods of *background correction* have been developed to eliminate these interferences. Here a basic description of these techniques is provided, along with their principal advantages and limitations. Further details on the theory and operation of background correction methods are provided in several reviews (64–67). Further information regarding the causes of background (scatter and molecular absorption) are given in Section 5.1.

The basic goal of background correction is the accurate measurement of the background absorbance. Subtraction of the background from the uncorrected signal (signal plus background) gives a background-corrected signal. An ideal method of background correction would be easy to use, inexpensive, and be applicable for all elements determinable by GFAAS. Also, it would be applicable for use with alternative wavelengths, in addition to the primary wavelengths, for all elements. Ideally, the background measurement would be made at the same wavelength as the analytical measurement to obtain best accuracy. The technique would use one light source and not affect source lifetime. No degradation of the detection limit or the linear range of calibration graphs would be observed. A technique that is compatible with a high sampling rate (>60 Hz) is necessary to accurately sample the transient GFAAS signal in which both analyte and background absorbance may vary with time. Finally, an ideal background correction method would also be well characterized by a wide variety of applications in the literature.

4.6.1 Nearby Line (Two-Line) Method

One of the first background correction methods employed for AAS involved the use of a second line, emitted by a source, that is nearby the analytical line and is not absorbable by the analyte. The second line may be an analyte line that cannot be absorbed by the atoms, a fill gas line in the hollow cathode lamp, or from another element in the same lamp (a multielemental HCL) or in a second lamp. This technique is called *nearby line*, or two-line, *background correction*. The nearby line must be close in wavelength to the analytical wavelength because the scatter signal is dependent on wavelength. The background at the nearby line must have a measurable and constant relationship to the background at the analytical line. This relationship must be unaffected by the sample matrix in order to obtain good accuracy. A list of recommended nearby lines for commonly determined elements is given in Table 4.2.

In practice, these criteria are difficult to meet. With the development of the modern background corrections described below, nearby line correction has

Table 4.2 Analytical and Background Lines for Nearby Line Background Correction (67)

Analyte	Analytical Wavelength, nm	Background Wavelength, nm	
		Analyte Lamp	Alternative Lamp Wavelength
Aluminum	309.3	307	Indium, 305.7
Antimony	217.6	217.9	—
Arsenic	193.7	192.0	—
Cadmium	228.8	226.5	—
Chromium	357.9	352.0	—
Cobalt	240.7	238.8	—
Copper	324.7	323.4	—
Iron	248.3	247.2	—
Lead	283.3	280.2, 287.3	—
Magnesium	285.2	281.7	Lead, 287.3
Manganese	279.5	—	Lead, 280.2
Nickel	232.0	231.6	—
Silver	328.1	326.2	—
Tin	286.3	283.9	—
Zinc	213.9	212.5	—

found little use in GFAAS. A notable exception is a commercially available, portable GFAAS instrument for lead determination designed by Sanford and co-workers (68). A lead HCL was the light source; a tungsten coil, from a halogen bulb of a photoprojector served as the atom cell; a fiber-optic cable and a charged-coupled device (CCD) were the detection system. The tungsten coil was housed in a glass cell and purged with 10% H_2/Ar. The hydrogen was introduced to prevent oxidation of the tungsten coil. A car battery may be employed as the power source for atomizer and instrument control was provided by a laptop computer.

Sanford and co-workers (68) reported that a major drawback of the original nearby line method was the use of sequential analytical and background measurements that were separated in time by minutes. The use of the CCD allowed simultaneous analytical and background measurements. The optimum accuracy was achieved by use of an analytical measurement at 283.3 nm and a background measurement obtained by the average of lines at 280.2 and 287.3 nm. Good accuracy was reported for the determination of lead in paint and blood standard reference materials. Although these preliminary results are encouraging, clearly more analyses need to be performed to assess the analytical utility of this system.

4.6.2 Continuum Source Method

Continuum source background correction was pioneered in 1965 for flame AAS by Koirtyohann and Pickett (69) and has been available commercially since around 1968. It is the most commonly used method of background correction. The most widely used continuum sources are the deuterium arc for metals that absorb in the UV range of 180 to 350 nm and a tungsten halide lamp in the visible range of 350 to 800 nm. A hydrogen lamp may be substituted for the deuterium lamp. Many commercial systems are equipped only with the deuterium arc because background levels are higher in the UV region.

The continuum source is aligned in the optical path of a spectrometer so that light from the continuum source and light from a primary source such as the hollow cathode lamp are transmitted alternately through the graphite furnace (Fig. 4.19). The spectroscopic principles of continuum source background correction are shown in Figure 4.20. The hollow cathode lamp has a narrow-band emission profile ($\sim$0.003 nm) (Fig. 4.20*a*), which may be attenuated by scatter, background, and the analyte. The continuum source has a relatively broad emission profile (Fig. 4.20*b*), which is defined by the spectral bandpass of the monochromator, typically 0.2 to 1 nm, and hence the amount of continuum source light absorbed by the analyte is negligible ($<$1%). However, the background, either broad molecular bands or scatter, can absorb the broad-band emission. In summary, analyte signal and background are measured by the HCL, while background is measured by the continuum source. The two measurements are subtracted electronically to give a background-corrected signal, ideally eliminating the effect of background signals.

Continuum source background correction generally accurately corrects for the relatively small background levels present in flame AAS. However, a number of GFAAS interferences have been reported for which a continuum source cannot accurately correct. One reason for inaccurate analyses is the

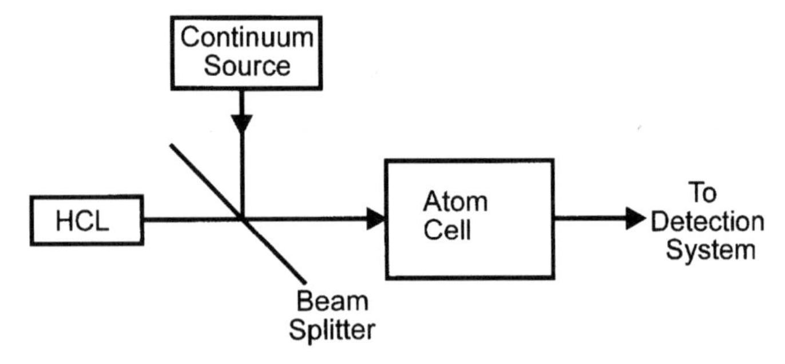

Figure 4.19. Schematic diagram of the instrumentation for continuum source background correction.

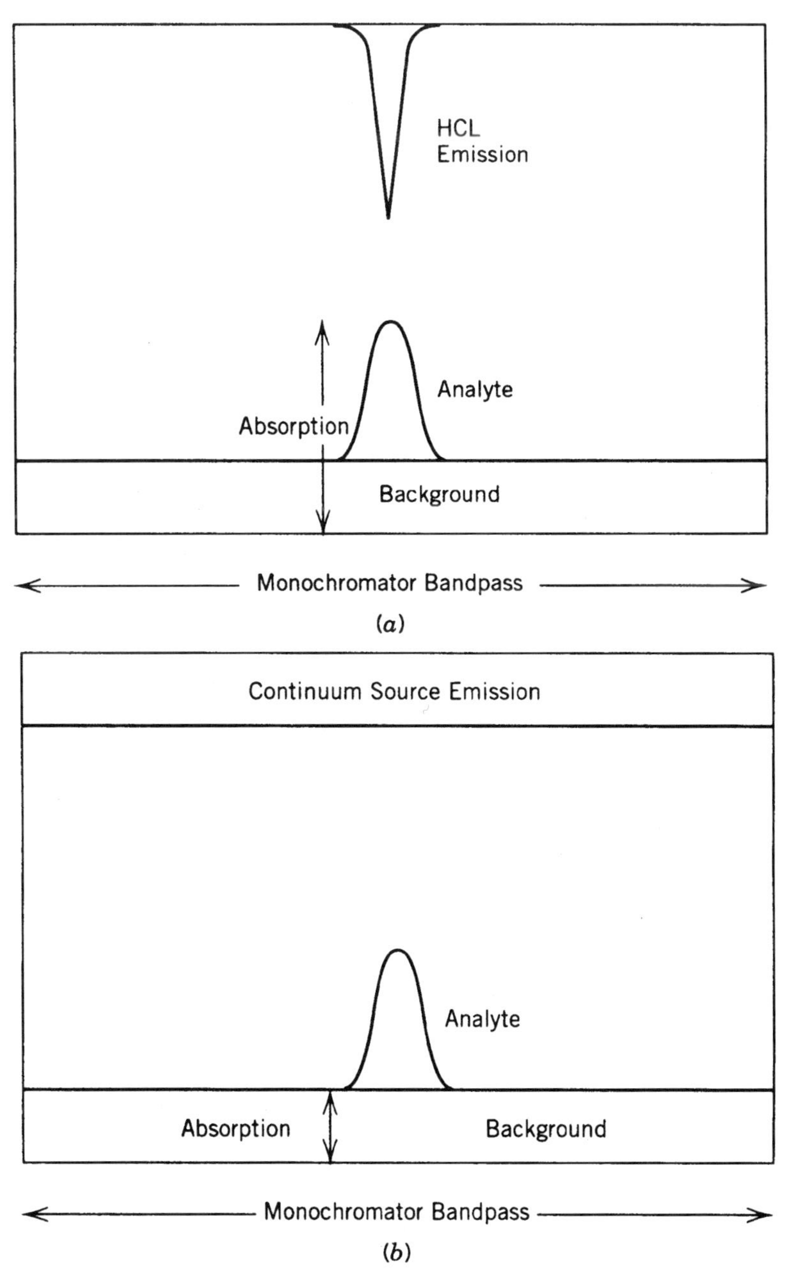

Figure 4.20. Schematic diagram of the basic principles of continuum source background correction: (*a*) hollow cathode lamp measurement (signal plus background) and (*b*) continuum source measurement (background). Subtraction of (*a*) minus (*b*) gives a background corrected signal.

impossibility of exactly aligning the HCL and continuum source beams, which implies that the accuracy of correction degrades as the background level increases. Although accurate correction has been demonstrated at background absorbance values up to 1.5 s, 0.5 s is generally considered the maximum level for which a continuum source can accurately correct (66,67).

Its second major limitation is its inability to correct for backgrounds, which vary with wavelength, across the bandpass of the monochromator, which are referred to as *structured backgrounds* (Fig. 4.21). This background correction method measures the average background across the spectral bandpass. Hence, if structured background is present, the actual background level may be higher than the average level determined by the continuum source, resulting in a corrected signal that is inaccurately high (undercorrection), or the actual level may be lower than that measured by the background-correction system, resulting in a corrected signal that is inaccurately low (overcorrection). Overcorrection is generally characterized by a negative dip in the corrected signal, as shown in Figure 4.22. If the structured background is relatively broad, it may be possible to improve accuracy by use of a narrower bandpass. Undercorrection is a more serious problem because it is difficult to diagnose. Traditional methods to verify accuracy, such as the method of standard additions (Section 3.3), will not correct

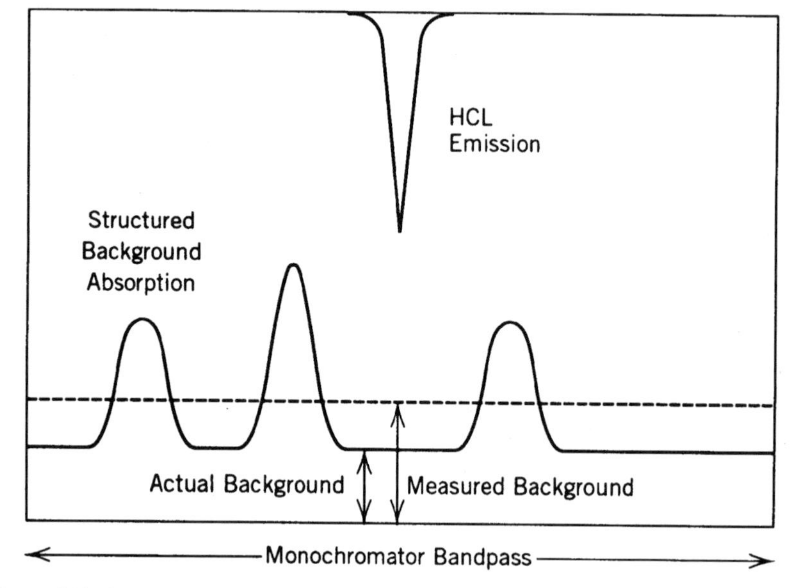

Figure 4.21. Schematic diagram of structured background with continuum source background correction. Notice that the background does not equal the continuum source background measurement.

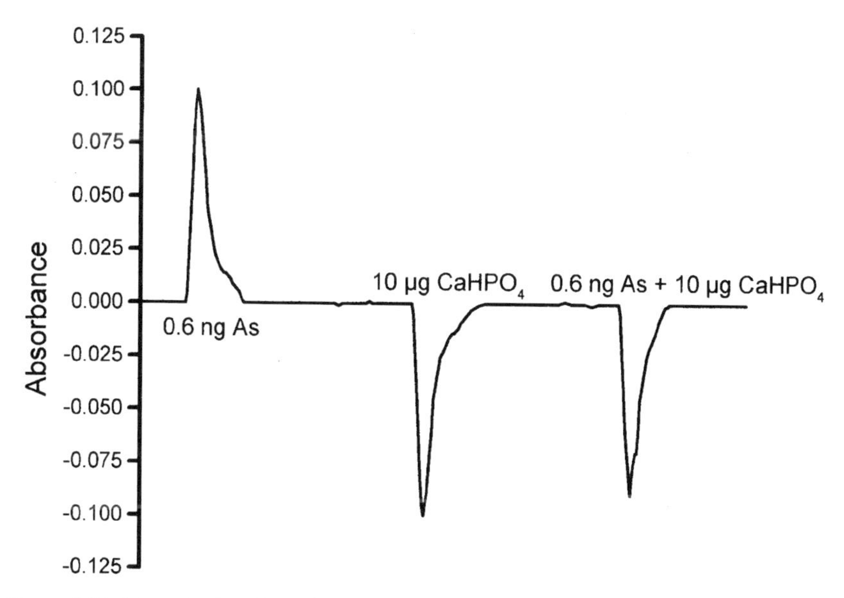

Figure 4.22. Determination of arsenic (193.7 nm) in the presence of $CaHPO_4$ with continuum source background correction and a 1-nm spectral bandpass (67).

for spectral interferences. These interferences can only be diagnosed by the use of an alternative wavelength or more powerful methods of background correction described later in this chapter. A list of potential continuum source background correction errors is given in Table 4.3.

In spite of these limitations for GFAAS, continuum source background correction has been used for a large number of analyses with good accuracy. Advantages include relatively low cost, applicability to all absorption lines accessible by the continuum source, and its suitability for use at higher sampling frequencies (up to 200 Hz) than other techniques. Relatively little degradation in detection limit or linear dynamic range is usually observed. Continuum source systems are often present in modern instruments alone or in conjunction with other background correction systems.

4.6.3 Self-Reversal (Smith–Hieftje) Method

Self-reversal background correction uses the phenomenon of self-reversal of a hollow cathode lamp at high lamp currents to measure the background absorbance (Section 4.1.1). Commercial development of the technique ensued after initial studies performed by Smith and Hiefjte (70), and hence the method is also called Smith–Hieftje background correction. The emission of an HCL

Table 4.3 Continuum Source Background Correction Interferences (105)

Analyte, Wavelength, nm	Interference, Wavelength, nm
As, 193.7	Phosphate
	Hair and milk
	P_2
	Fish tissue
Au, 242.8	Co, 242.5
Bi, 223.1	Cu, 223.0
	Co
Cd, 228.8	Natural waters
	Fish tissue
	Wastewater
Ga, 287.4	Fe, 287.4
Hg, 253.6	Co, 253.6
Ir, 285.0	Mg, 285.2
P, 213.6	Fe, 213.6
	Steel
Pb, 217.0	Cu, 216.5
	Sb, 217.0
Pb, 283.3	Wastewater
Pt, 271.9	Fe, 271.9
Sb, 217.6	Co, 217.5
	Fe, 217.8
	Cu, 217.9
	Pb, 217.0
	Gunshot residue
Sb, 231.15	Ni, 231.1
Se, 196.0	Fe
	Blood
	Phosphate, urine
	Ni alloys, Fe
	Fish tissue
	Co
Se, 204.0	Cr
	Ni
Te, 214.3	Phosphate
	Zn, 213.9
Zn, 213.9	Te, 214.3
	Fe, 213.9

operated at normal low current (0–30 mA) is absorbable by the analyte and the background (Fig. 4.23*a*). However, as the lamp current is increased to high values (300 mA), the amount of light absorbed by the atoms decreases. The emission from a lamp operated at high current will not be significantly absorbed by the analyte because the emission profile is sufficiently self-reversed to prevent absorption in the ideal case or at least significantly reduce it, which is generally what actually occurs (Fig. 4.23*b*). However, the background is broad-band compared to the source emission, and its absorbance is unaffected by self-reversal. Hence background correction is accomplished by alternate measurement of the low current (signal and background) and high current (background). Subtraction of these two measurements gives the background-corrected signal. The signal plus background measurement is made at low current for several milliseconds, while the background measurement is made at high current for several hundred microseconds. The measurement times are precisely controlled so that the same number of photons reach the detector during each phase of the cycle.

This technique for background correction has been developed commercially by Thermo Jarrell Ash (Franklin, MA), Shimadzu Scientific Instruments, Inc. (Columbia, MD), Buck Scientific (East Norwalk, CT), and Leeman Laboratories (Lowell, MA). Advantages include low cost, the use of one source (no problems superimposing two sources), and measurement of the background signal a few picometers away from the analytical wavelength. The two last features allow correction at higher absorbance values and more accurate analysis compared to continuum source methods. A classic example of this is illustrated in Figures 4.24 (self-reversal) and 4.22 (continuum source) for the determination of arsenic in the presence of $CaHPO_4$. Although severe overcorrection was observed with the continuum source system, good accuracy was obtained with the self-reversal technique.

Several limitations are associated with this technique. First, self-reversal background correction cannot be used with many involatile elements because their HCLs cannot be sufficiently self-reversed. Other refractory elements, such as silicon and titanium, have sensitivity losses of 80% by self-reversal compared to the uncorrected signal. Moderately involatile elements (aluminum, chromium, platinum) can be determined by this technique, but because relatively little self-reversal is observed for these elements, the background-corrected signal is reduced by 50%, causing a degradation in characteristic mass values.

Virtually all elements show a reduction in linear dynamic range (LDR) (66,67). This reduction occurs because stray light determines the limiting absorbance. Some elements, such as copper, have greater levels of stray light at high current than low current, which implies that the limiting absorbance value for the background measurement is smaller than for the uncorrected signal. For these elements, a small dip in the absorbance profile is obtained, inducing a

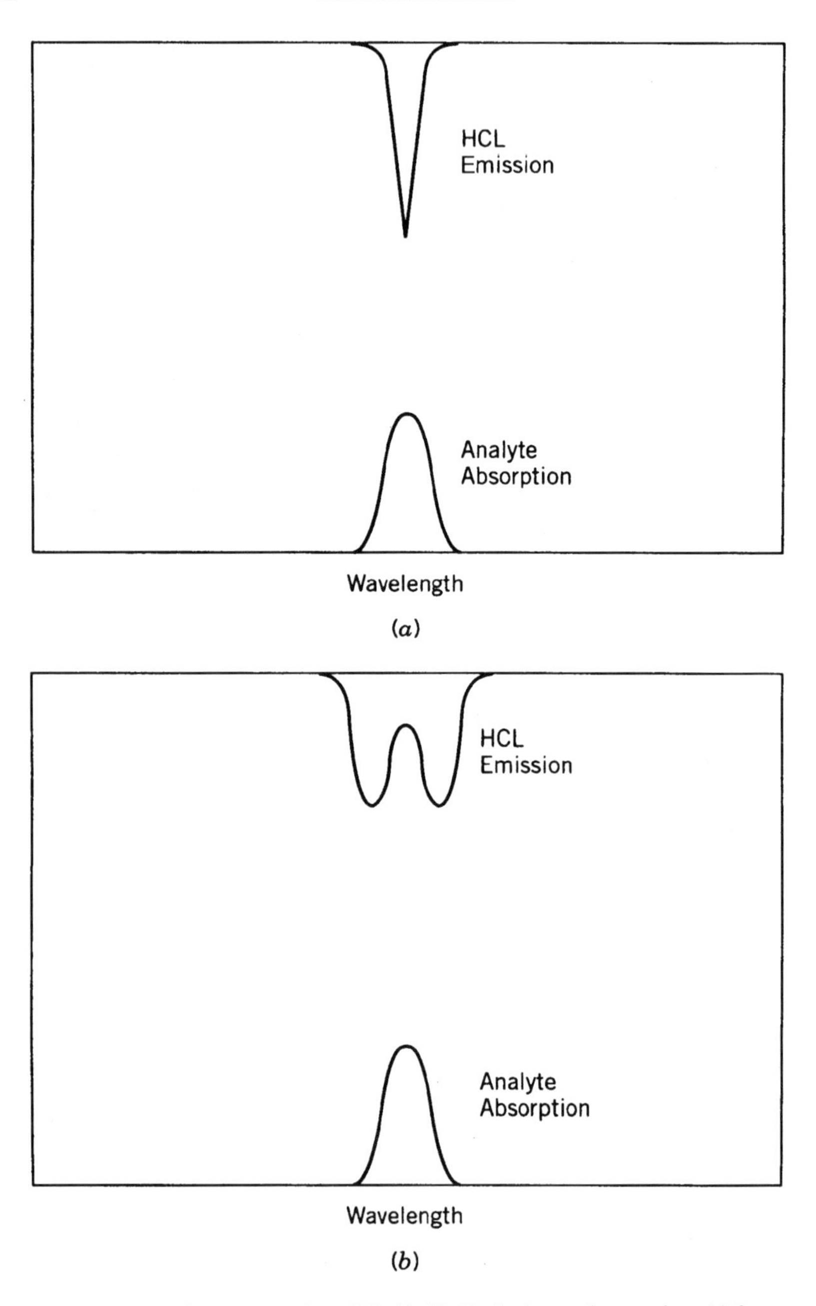

Figure 4.23. Schematic representation of Smith–Hieftje background correction: (*a*) low current measurement of signal and background and (*b*) high current measurement of background. Subtraction of (*a*) minus (*b*) gives a background-corrected signal.

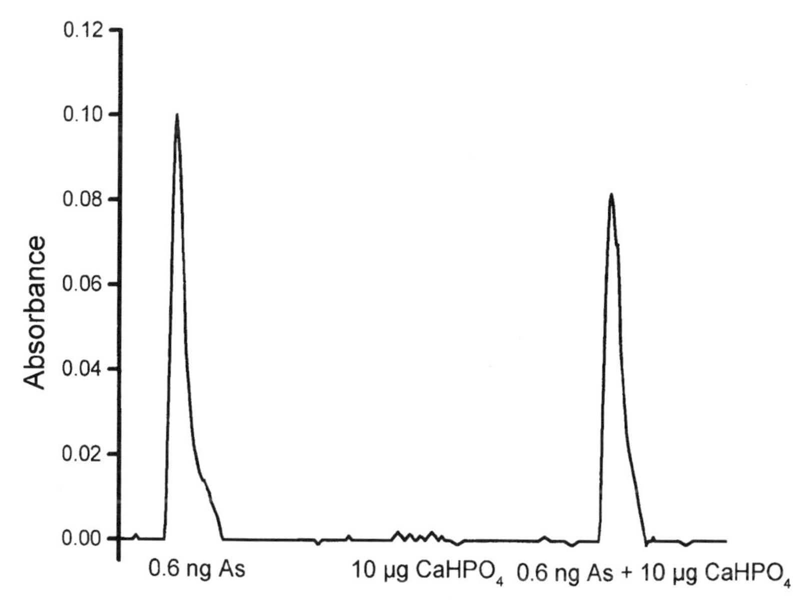

Figure 4.24. Determination of arsenic in calcium phosphate matrix by GFAAS with Smith–Hieftje background correction systems (67).

decrease in the slope of the calibration curve at high concentrations, but no rollover.

In general, rollover is not a severe problem with volatile elements because it occurs at relatively high concentrations. However, for involatile elements, whose HCL emission profiles are not well self-reversed, the LDR can be degraded to one order of magnitude or less.

Self-reversal is limited to resonance transitions because only transitions originating from the ground state show self-reversal. Self-reversal correction cannot be used with EDLs, which are commonly used for the determination of several volatile elements (e.g., arsenic, lead, selenium). Special HCLs, which are more expensive than the conventional lamps, must be employed with this method. In general, their lifetime is reduced by one-third compared to conventional HCLs. However, this system has the advantage that an additional (continuum) source is not required with its own operating costs and finite lifetime.

4.6.4 Zeeman Effect Method

Zeeman discovered that an intense magnetic field causes atomic energy levels to split, causing atomic lines to split into two or more components (71,72). These components generally only absorb one type of polarized light. Consequently, the combination of a magnetic field with a polarizer may be used to prevent absorption by the analyte at the analytical wavelength, allowing measurement of background. The pattern of components observed depends on the element, the spectroscopic transitions employed, and the orientation of the magnetic field with respect to the radiation.

The case where the magnetic field is oriented perpendicular to the HCL light is called *transverse Zeeman* (Fig. 4.25*a*, 4.25*b*). The simplest splitting pattern is observed for elements that demonstrate the *normal Zeeman effect* (Fig. 4.26). An element that undergoes the normal Zeeman effect in a transverse field is split into three components, one π component at the analytical wavelength, which can only absorb light polarized in the plane parallel to the magnetic field, and two σ components, which are arranged equal distances on opposite sides of the π component, and can only absorb light polarized in the plane perpendicular to the magnetic field. The relative intensity of the triplet is 1 : 2 : 1. The distance that the σ components are shifted from the π component is proportional to the strength of the magnetic field.

Alternatively, the magnetic field may be oriented parallel to the radiation, which is called *longitudinal Zeeman* (Fig. 4.25*c*, 4.25*d*). Splitting observed in normal longitudinal Zeeman consists of two σ components, each shifted an equal distance from the original wavelength, without any π component (Fig. 4.27). The σ components each absorb one of the two directions of circularly polarized light.

The transitions of most elements undergo a more complicated pattern of splitting than the normal Zeeman effect, called the *anomalous Zeeman effect*, consisting of several π and σ components. An example of an anomalous Zeeman pattern is shown in Figure 4.28.

In practice, the magnetic field may be applied to the light source, which is called *direct Zeeman*, or to the atom cell, which is called *inverse Zeeman* (Fig. 4.25). A polarizer (either stationary or rotating) is used to remove the π component in a transverse magnetic field to make the background measurement. Finally, either a permanent (dc) magnet or an electromagnet (ac) may be used, with typical magnetic fields between 5 and 15 kG.

Zeeman effect background correction may be used with virtually all elements and transitions determined by GFAAS but requires a relatively expensive magnet and electronics for synchronization with the other components of the instrument. From the discussion above, it is clear that a wide variety of configurations for Zeeman effect AAS are possible. Zeeman effect background correction was developed for GFAAS in the 1970s and has gained prominence with the introduction of commercial systems by many leading manufacturers of graphite furnace AAS instrumentation including Perkin-Elmer (Norwalk, CT), Varian Associates (Palo Alto, CA), Hitachi Instruments (Danvers, CT), and

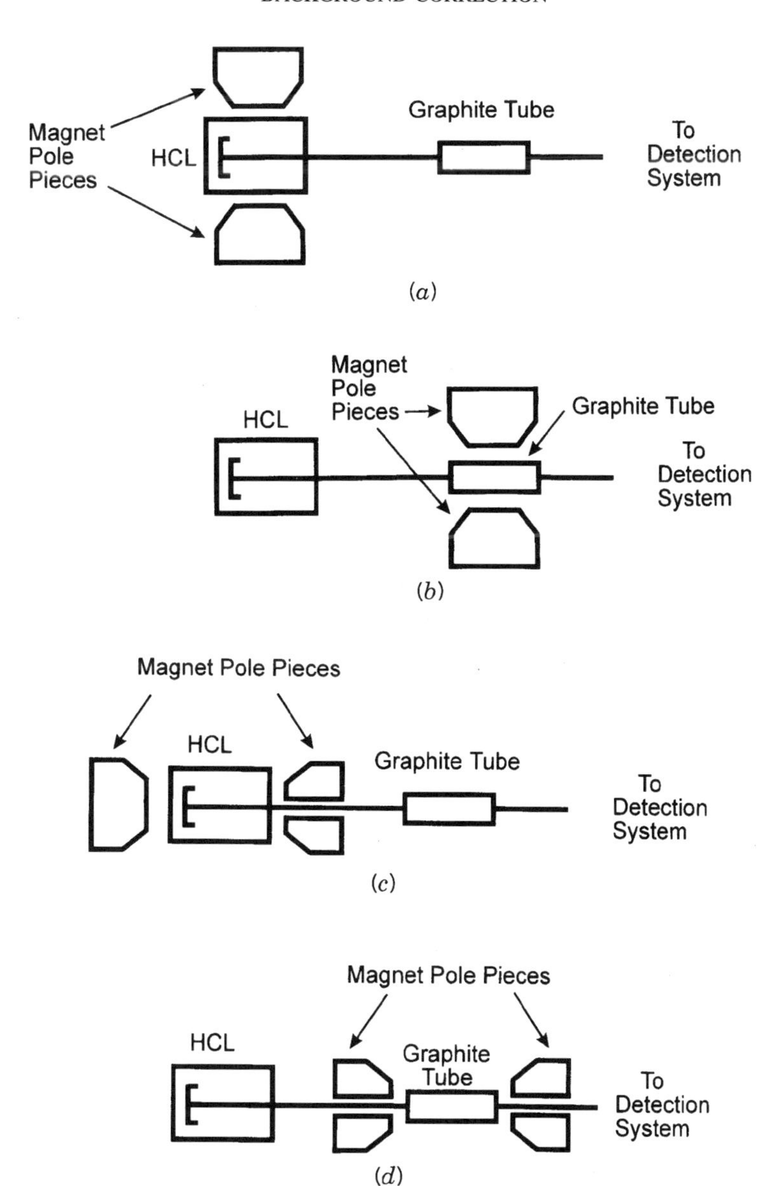

Figure 4.25. Configurations for Zeeman background correction: (*a*) direct, transverse Zeeman; (*b*) inverse, transverse Zeeman; (*c*) direct, longitudinal Zeeman; and (*d*) inverse, longitudinal Zeeman.

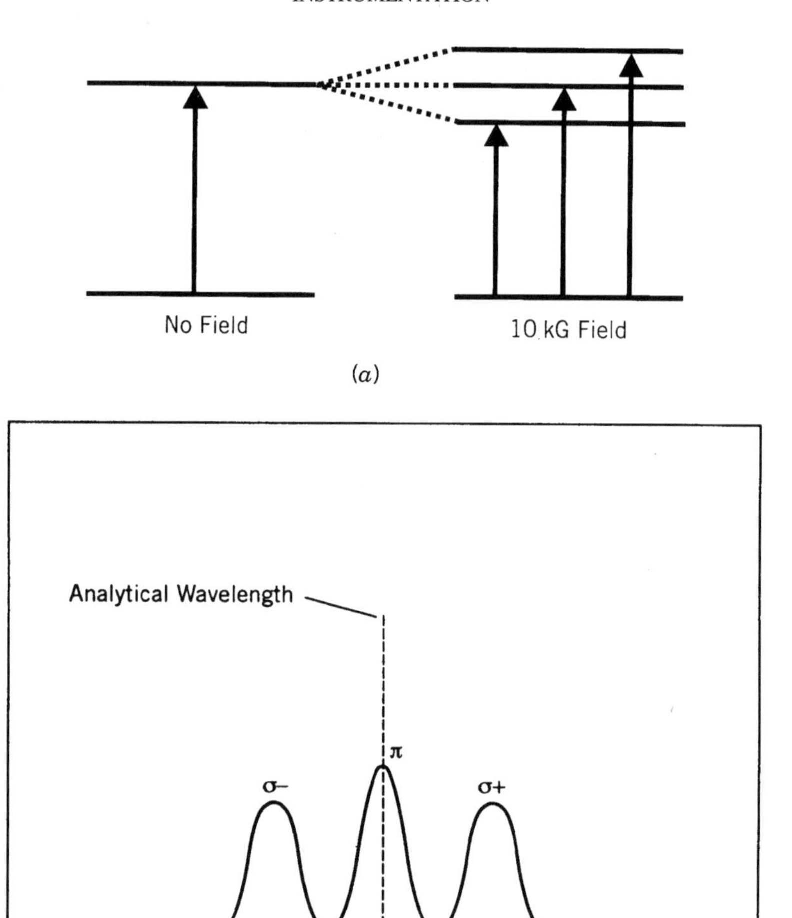

Figure 4.26. Normal Zeeman splitting pattern in a transverse magnetic field: (*a*) spectroscopic transitions and (*b*) wavelength profile.

Thermo Jarrell Ash–Unicam (Franklin, MA). Here, three inverse configurations commonly employed today in commercial instrumentation are discussed: inverse longitudinal ac Zeeman, inverse transverse ac Zeeman, and inverse transverse dc Zeeman (Fig. 4.29). Direct Zeeman has not been widely used because of the difficulties of operating a light source in a strong magnetic field, and consequently, direct Zeeman is not discussed here. The reader is referred to reviews by de Loos-Vollebregt and de Galan (71) and Yasuda et al. (72) for

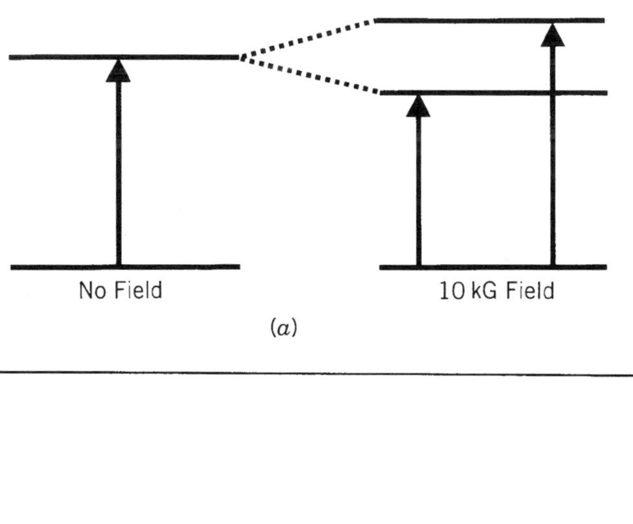

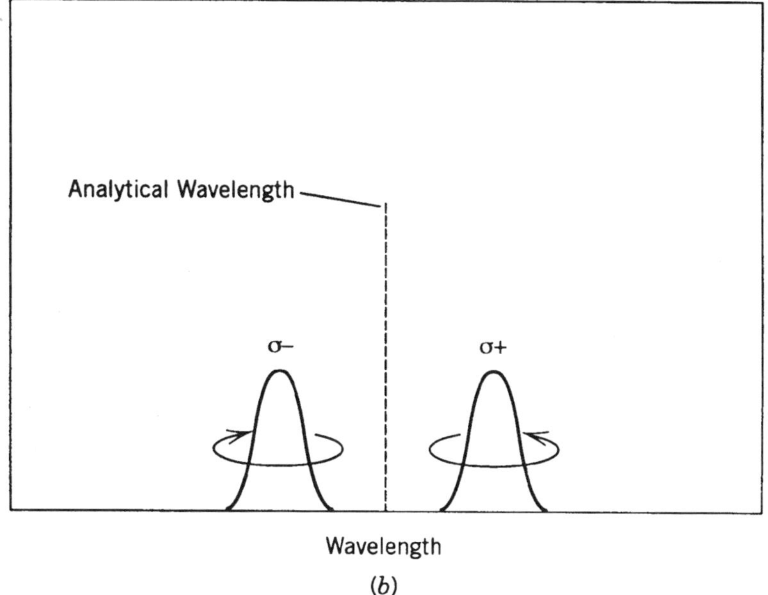

Figure 4.27. Normal Zeeman splitting pattern in a longitudinal magnetic field: (*a*) spectroscopic transitions and (*b*) wavelength profile.

further discussion on theory and other instrumental configurations of the Zeeman effect for GFAAS.

4.6.4.1 Inverse Longitudinal ac Zeeman

This configuration was developed in the late 1980s and early 1990s (73,74) to facilitate the use of Zeeman effect background correction with transversely heated graphite furnaces (Section 4.2.1) and is available commercially from

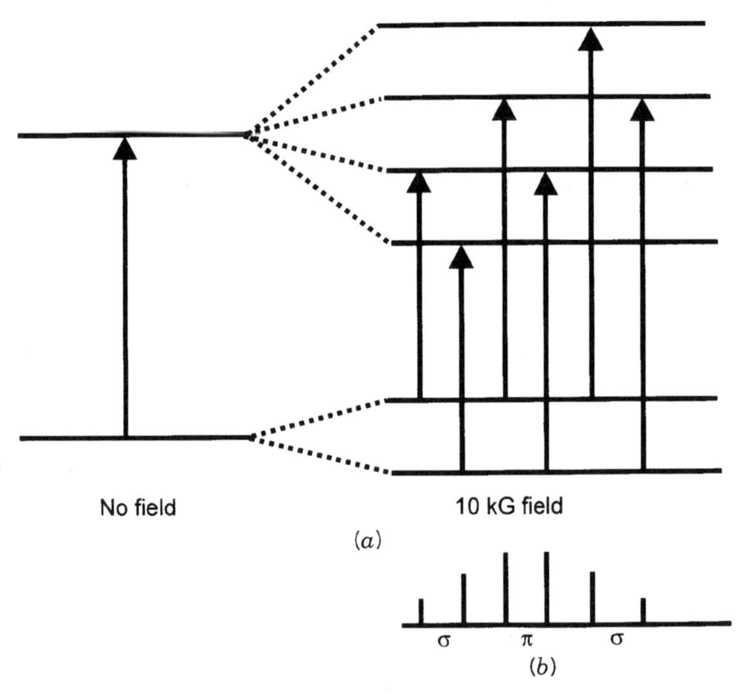

Figure 4.28. Example of an anomalous Zeeman pattern: (*a*) spectroscopic transitions and (*b*) wavelength profile.

Perkin-Elmer (Norwalk, CT). Its relatively late development is attributed to the difficulty of engineering a homogeneous field while providing holes for the HCL beam to pass. Once the engineering problems were solved, this design is probably the optimum Zeeman configuration. The line frequency of 50 (Europe) or 60 (United States) Hz is used to power the magnet. The line frequency is rectified to allow a duty cycle of 100 or 120 Hz, which is synchronized with the light source, which is sufficiently fast to allow accurate sampling of the transient signals.

The mechanism of background configuration is shown in Figure 4.30. When the magnetic field is off, HCL light is absorbed by the analyte and the background. When the magnetic field is on, the atomic lines in the atom cell are split into σ components located away from the analytical wavelength, while the background is (ideally) unaffected. The magnetic field-off measurement minus the field-on measurement gives a background-corrected signal.

Examination of Figure 4.30 shows that no π component is present, and hence a polarizer is not necessary to prevent analyte absorption during the background measurement. The absence of a polarizer is an advantage because no attenuation

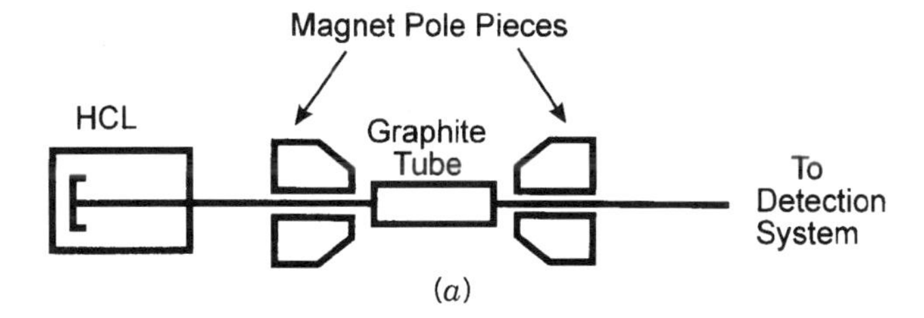

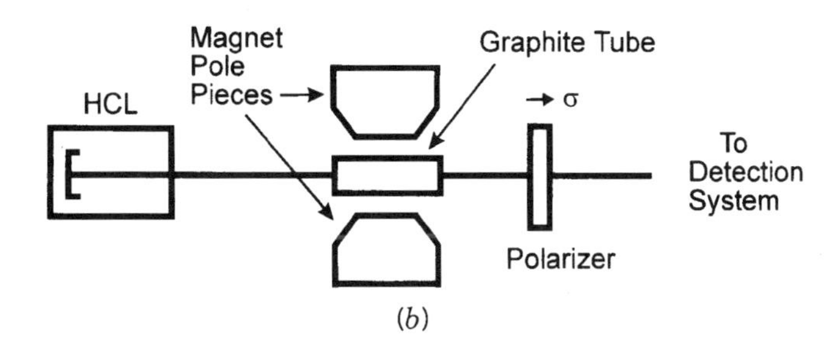

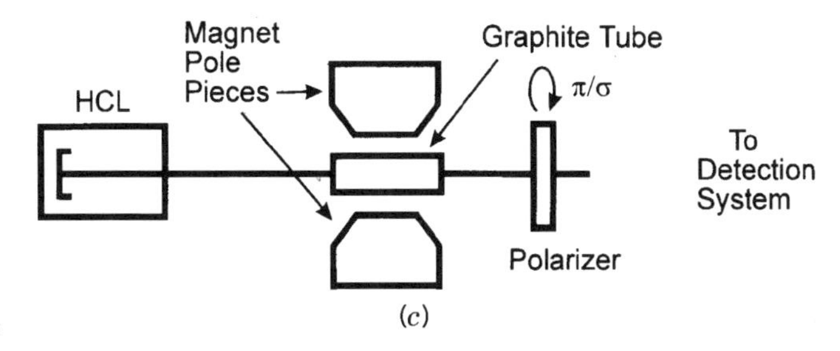

Figure 4.29. Inverse Zeeman AAS configurations: (*a*) inverse longitudinal ac Zeeman, (*b*) inverse transverse ac Zeeman, and (*c*) inverse transverse dc Zeeman.

of the source occurs, improving the signal-to-noise ratio (Section 4.1.1). The background correction measurement is made at the analytical wavelength, and hence this configuration will accurately correct for structured background, as long as the background levels are the same in the field-off and field-on measurements. This is always the case for scatter and has been shown to be true for the absorption spectra of most molecules. A few reports have demonstrated interferences due to the presence of concomitant element or a molecule with an absorption line within 10 pm of the analytical wavelength. The best approach to handle these relatively infrequent interferences is the use of an alternative wavelength. A list of some reported Zeeman interferences is given in Table 4.4.

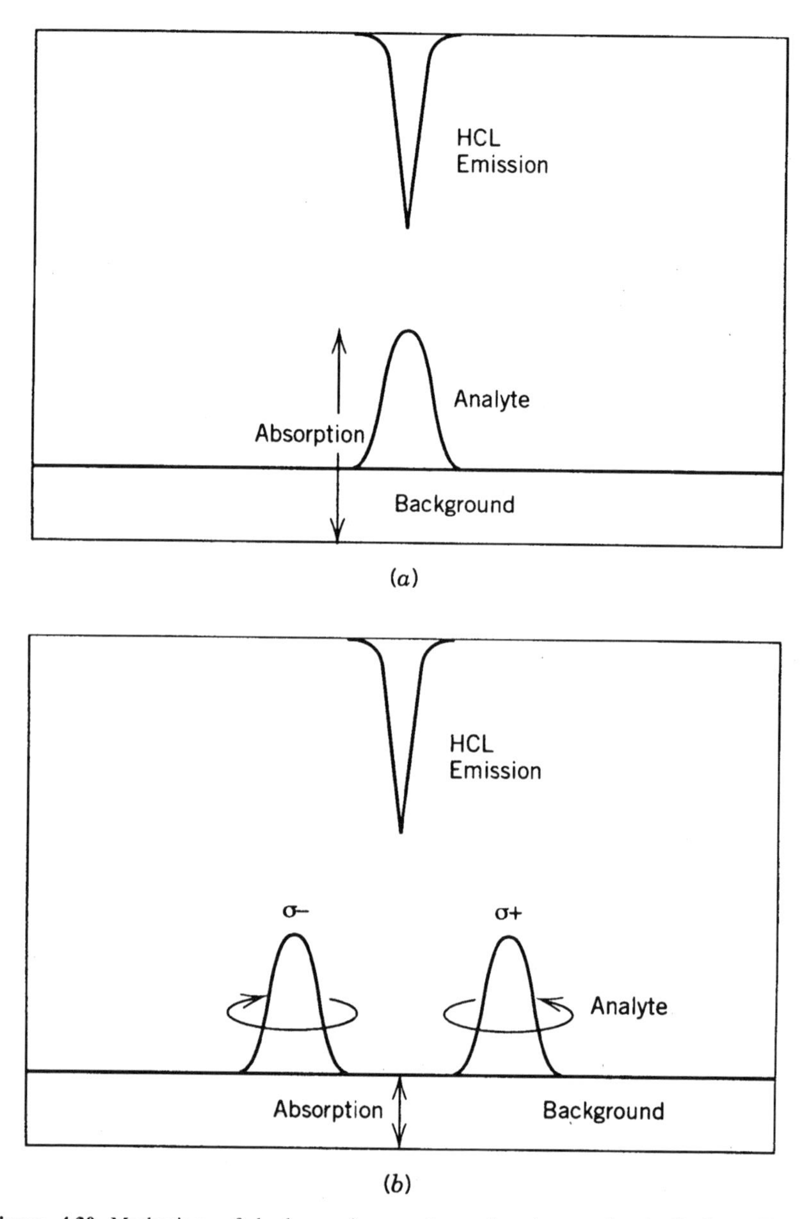

Figure 4.30. Mechanism of background correction using inverse longitudinal ac Zeeman background correction: (*a*) magnetic field off-measurement of analyte and background absorption and (*b*) magnetic field on-measurement of background absorption. Subtraction of (*a*) minus (*b*) gives a background-corrected signal.

Table 4.4 Zeeman Effect Background Correction Interferences

Analyte, Wavelength, nm	Interferent, Wavelength, nm	References
Ag, 328.068	Rh	106
Ag, 328.068	PO	66, 107
Al, 308.215	V, 308.211	66
Au, 267.595	Co, 267.598	66
Au, 267.595	BaO	106
B, 249.678	Co, 249.671	66
Bi, 227.658	Co, 227.653	66
Cd, 228.8	PO	76
Cd, 326.106	PO	66, 107, 108
Co, 243.666	Pt, 243.669	66
Cr, 357.9	CN	109
Cr, 360.5	Co	110
Cu, 244.2	PO	108
Cr, 260.553	Co, 360.535	66
Eu, 459.402	V, 459.411	66
Fe, 246.3	PO	107
Fe, 247.3	PO	108
Fe, 271.902	Pt, 271.904	66
Ga, 287.424	Fe, 287.417	66
Hg, 253.652	Co, 253.649	66, 107
In, 325.8	PO	108
Mn, 279.5	AlBr, InBr	106
Ni, 232.0	PO	76
Ni, 341.477	Co, 341.474	66
Ni, 305.082	V, 305.089	66
Pb, 283.3	PO	75, 76
Pb, 283.3	S_2	111
Pb, 283.3	AlBr, InBr, BaO	106
Pb, 217.0	AlO	106
Pb, 261.418	Co, 261.413	66
Pd, 247.642	Pb, 247.638	66
Pd, 247.642	PO	108
Pd, 244.791	PO	108
Pt, 265.945	Eu, 265.942	66
Pt, 306.471	Ni, 306.462	66
Pt, 272.396	Fe, 273.400	66
Si, 250.690	Co, 250.688	66
Si, 250.690	V, 250.691	66
Sn, 286.3	InBr, BaO, MgBr	106
Sn, 303.412	Cr, 303.419	66
Sn, 300.915	Ca, 300.921	66
Zn, 213.856	Fe, 213.859	66

Examination of Table 4.4 shows a number of interferences caused by PO, which is produced from phosphate chemical modifiers (75). It may be advisable to avoid the use of phosphate modifiers whenever possible to prevent these interferences. Heitmann et al. (76) made high-resolution measurements of a magnesium nitrate ammonium dihydrogen phosphate modifier across a 0.1-nm window across the cadmium 228.8-nm resonance line with and without a magnetic field (Fig. 4.31). Structured background was observed that was attributed to PO. The magnetic field induced significant changes to the absorption spectrum, with differences in integrated absorbance of ±0.2 s observed across the window (Fig. 4.31). These data indicate that Zeeman effect background correction may not provide accurate results when an absorption line is surrounded by structured background.

A reduction in the analyte signal is observed in this configuration due to absorption of source light by the σ components during the background measurement. This sensitivity reduction is less than 20% for most elements and less than 50% for almost all commonly determined elements, causing a small degradation in characteristic mass (less than a factor of 2). Figure 4.32 shows

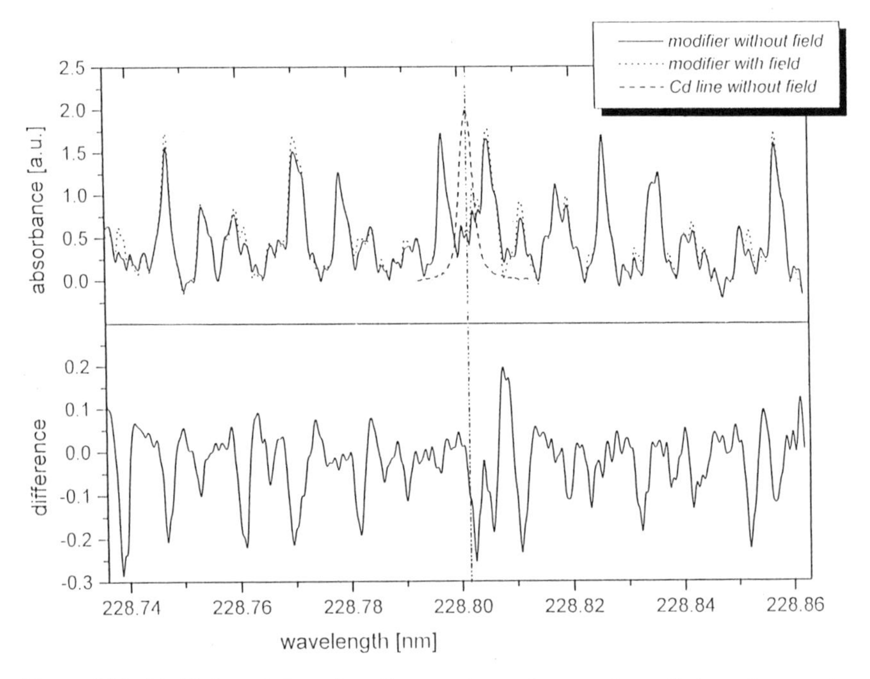

Figure 4.31. (*a*) High-resolution absorption spectrum of magnesium nitrate and ammonium dihydrogen phosphate with and without magnetic field in comparison to the cadmium line profile and (*b*) difference of field-off and field-on absorption spectra. Taken with permission from Reference 76.

that the uncorrected signal reaches a plateau before the background, and hence a reduction in the linear dynamic range and the presence of a maximum in the calibration curve are observed. The latter phenomenon implies that one absorbance value may correspond to two concentrations. Mathematical algorithms have been developed that produce integrated absorbance values corrected for these effects (77–81). These *linearization models* allow extension of calibration curves by a factor of 3 to 5 for many elements.

4.6.4.2 Inverse Transverse ac Zeeman

The instrumentation and basic principles of this Zeeman effect configuration are shown in Figures 4.29*b* and 4.33. The engineering required for this design is less exacting than the longitudinal ac Zeeman, and hence it was first introduced in the early 1980s and is currently commercially available from Perkin-Elmer (Norwalk, CT), Varian Associates (Palo Alto, CA), and Thermo Jarrell Ash–Unicam (Franklin, MA). The field-off measurement provides measurement of signal and background absorption, the field-on measurement gives the background absorption, and subtraction of the two supplies the background-corrected signal. In general, the basic principles are the same as the longitudinal case, except a π component is present during the background measurement, and hence a stationary polarizer is used to remove parallel polarized radiation from the source. The removal of 50% of source radiation by a polarizer causes a small degradation in signal-to-noise ratio (Section 4.1.1). The background measure-

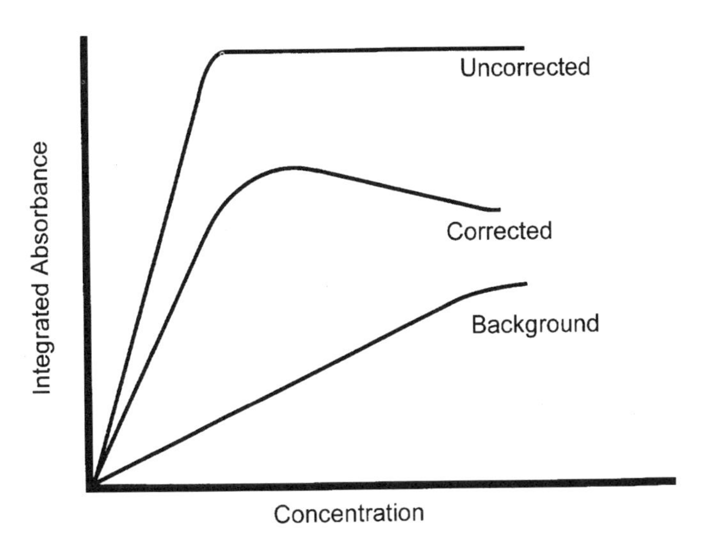

Figure 4.32. Schematic diagram of rollover in Zeeman AAS.

ment is made at the analytical wavelength, and in the absence of spectral overlaps, accurate correction is assured as long as the background absorbance is the same with the magnet off and on. The considerations with regard to sensitivity loss due to absorption by σ components are also the same as the longitudinal case.

A classic example of the ability of this Zeeman effect configuration to accurately correct for background is shown in Figure 4.34 for the determination of lead in seawater (82).

4.6.4.3 Inverse Transverse dc Zeeman

This configuration (Figs. 4.29*c* and 4.35) was incorporated in the first commercial Zeeman effect GFAAS system developed by Hitachi Corporation (Danvers, CT) in the mid-1970s and is currently available from this manufacturer. A rotating fixed polarizing prism or dual refracting filter alternately passes parallel and perpendicular radiation from the light source. The π component and the background may absorb parallel light from the HCL, allowing the measurement of signal and background absorbance. When the polarizer is rotated to allow only the perpendicular radiation to pass, only the σ components, located away from the analytical wavelength, may absorb the source radiation. If the background is unaffected by the magnetic field (Section 4.6.4.1), as is usually the case, subtraction of the parallel measurement minus the perpendicular measurement gives a background-corrected signal. The polarizer is rotated at 100 Hz, which is sufficiently fast to allow accurate measurement of the analyte signal and background.

Like the other inverse Zeeman effect configurations, the dc system has the advantage of correction at the analytical wavelength. Additional complexity is introduced by the use of a rotating polarizer, although this is offset by the use of a permanent magnet instead of an electromagnet. The dc system provides similar sensitivity to the ac systems for elements that display the normal Zeeman effect. However, the dc system is less sensitive for anomalous Zeeman elements because the π components are split in the magnetic field (Fig. 4.28). A decrease in the magnetic field to increase the absorption by the π component induces greater absorption by the σ components during the background measurement, resulting in a sensitivity loss. The practical effect is that the sensitivity of dc configuration is degraded for anomalous Zeeman elements, with sensitivity reductions as high as 60 to 70% for some elements.

4.6.5 Comparison of Background Correction Systems

A comparison of commercially available GFAAS background correction methods is given in Table 4.5. The first major criterion is accessibility. The

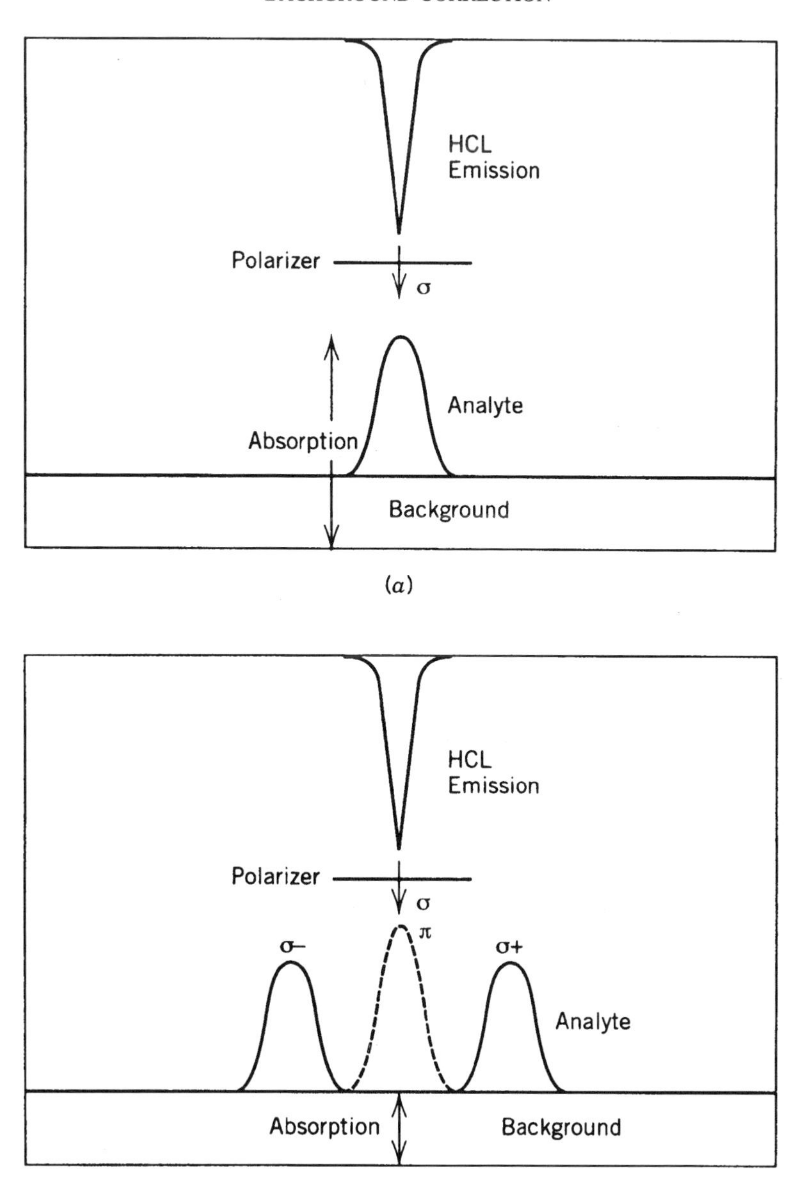

Figure 4.33. Mechanism of background correction using inverse transverse ac Zeeman background correction: (*a*) magnetic field off-measurement of analyte and background absorption and (*b*) magnetic field on-measurement of background absorption. Notice that the π component cannot absorb source light during the background measurement because of the polarization of the light. Subtraction of (*a*) minus (*b*) gives a background-corrected signal.

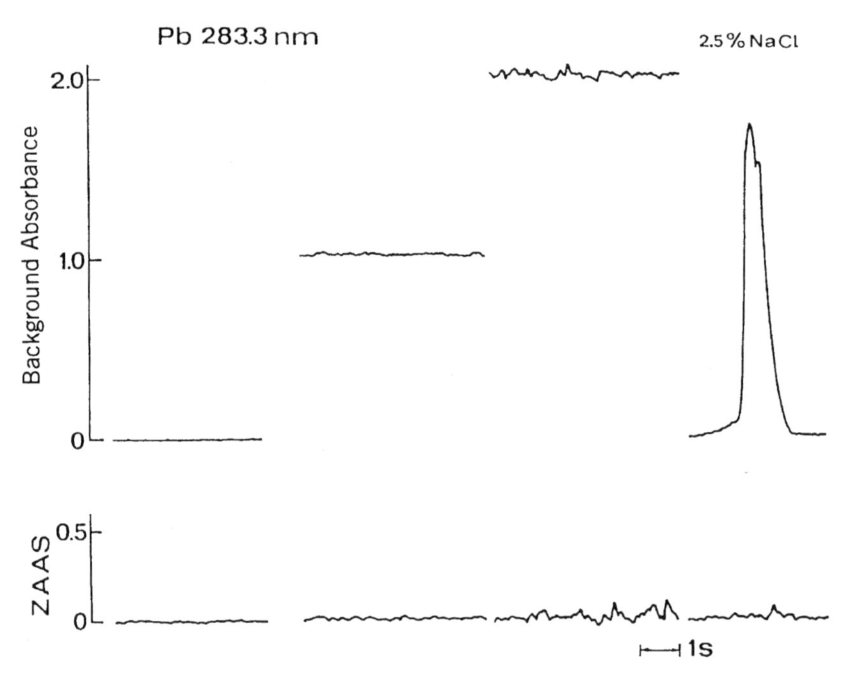

Figure 4.34. Determination of lead in seawater by GFAAS for uncorrected signal (in transmittance mode) and Zeeman-corrected signal (in absorbance mode). Taken with permission from Reference 82.

continuum source is limited by the availability of analyte lines that overlap its useful spectral region (for a deuterium arc, 190–350 nm), although this is not a problem for most elements. Self-reversal correction is limited to the resonance lines of relatively volatile elements, and hence many analysts find it convenient to have a continuum source system as well to do involatile elements and use alternative transitions. The Zeeman methods are generally accessible to all elements and transitions.

The next comparison involves the performance of a correction system to account for background. Continuum systems have been shown to have interferences due to the impossibility of superimposing the output of two sources and correcting for structured background. Self-reversal and Zeeman methods correct near (the former) or at (the latter) the analytical wavelength, and hence relatively few interferences have been reported.

The next set of criteria include analytical figures of merit. Minimal sensitivity loss (<2–3 times) is observed with continuum source, the ac Zeeman, and self-reversal for volatile elements. Significant degradation of detection limits is observed with self-reversal method with involatile elements and de Zeeman with

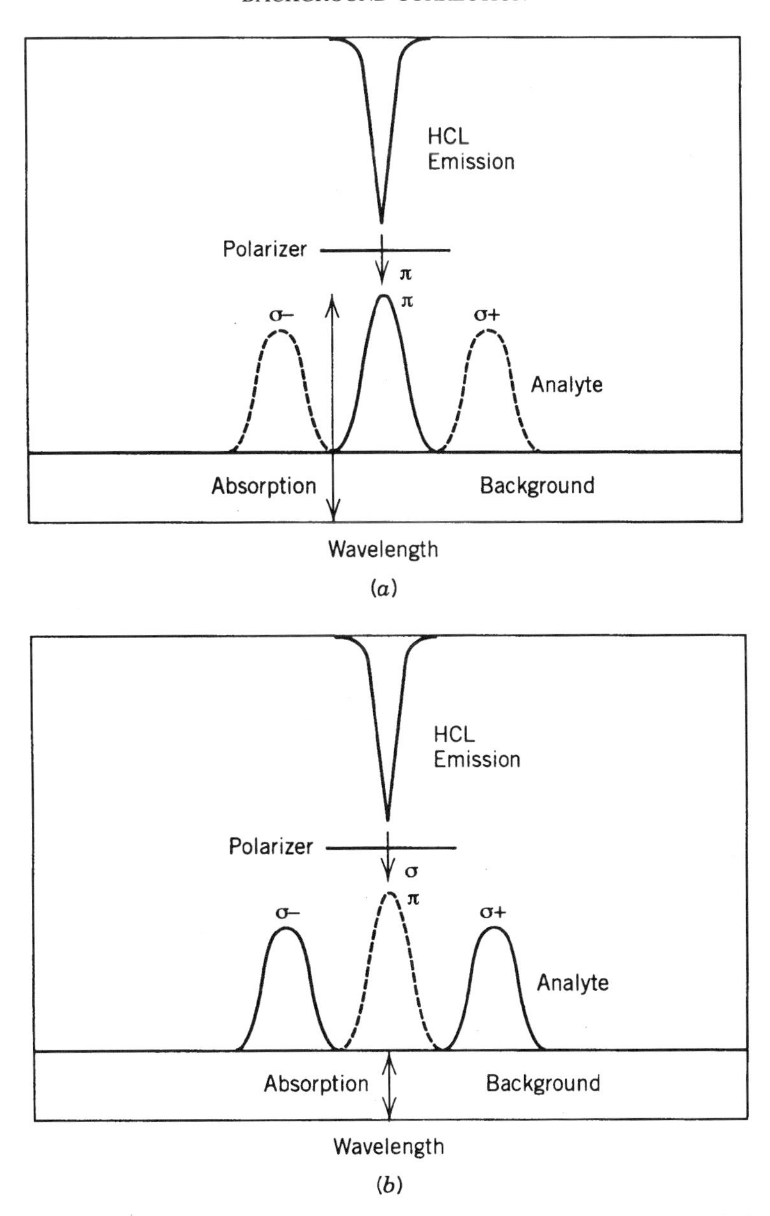

Figure 4.35. Mechanism of background correction using inverse transverse dc Zeeman background correction: (*a*) transmission of parallel polarized radiation, measurement of analyte and background absorption, and (*b*) transmission of perpendicular polarized radiation, measurement of background absorption. Notice that the π component cannot absorb source light during the background measurement because of the polarization of the light. Subtraction of (*a*) minus (*b*) gives a background-corrected signal.

Table 4.5 Comparison of Commercially Available Background Correction Methods in GFAAS

Criterion	Method of Background Correction				
			Inverse Zeeman		
	Continuum Source	Self-reversal	Longitudinal ac	Transverse ac	Transverse dc
Accessible elements	All wavelengths accessible with the continuum source	Not applicable for refractory elements	All	All	All
Accessible transitions	All wavelengths accessible with the continuum source	Resonance transitions	All	All	All
Location of background measurement	Across monochromator bandpass	3–5 pm from analytical wavelength	At analytical wavelength	At analytical wavelength	At analytical wavelength
Reported interferences	Many	Few	Few	Few	Few
Sensitivity loss	Minimal	Minimal for volatile elements, substantial for involatile elements	Minimal	Minimal	Substantial for anomalous Zeeman elements

Reduction of linearity of calibration curves	Minimal	Significant, particularly for involatile elements	Significant	Significant	Significant
Sampling frequency	100–200 Hz	20–100 Hz	100–120 Hz	100–120 Hz	100 Hz
Applications	Many	Few	Few (although interferences would be similar to ac transverse)	Many	Some
Additional instrumentation	Source and electronics	Electronics	Electromagnet and electronics	Electromagnet, polarizer, and electronics	Permanent magnet, rotating polarizer, and electronics
Capital cost	Low	Low	High	High	High
Operating costs	Additional source	Increased cost and reduced lifetime of HCLs	Power to operate magnet	Power to operate magnet	None
Ease of use	Relatively difficult; requires alignment of two sources	Easy	Easy	Easy	Easy

anomalous Zeeman elements. The continuum source method causes minimal reduction of the linear range of calibration curves compared to the other methods. Continuum source correction can be operated at higher sampling frequencies than the other methods (up to 200 Hz), although the sampling frequencies used for the Zeeman methods and self-reversal methods in most commercial instruments (100–120 Hz) are probably adequate. Continuum source and transverse ac Zeeman have the advantages of a large number of applications, although interferences would be expected to be the same with the transverse and longitudinal Zeeman systems.

Another concern is the cost of the various systems. Self-reversal systems are the least expensive, only requiring additional electronics, and have low capital costs. However, self-reversal requires the use of more expensive HCLs with a reduced lifetime. The capital costs of continuum systems are also low, although an additional source and associated electronics are required. The Zeeman systems require a magnet, which significantly adds to the cost of these systems, but the operating costs are the lowest.

In terms of ease of use, self-reversal and the Zeeman systems are preferable. They do not require alignment of two light sources required with continuum source systems.

In conclusion, we would recommend that most GFAAS instruments should be equipped with two background correction systems: continuum source and one of either Zeeman or self-reversal. A continuum system is useful for a large number of analyses and elements/transitions inaccessible by the other methods. Self-reversal has lower capital costs, although it is more expensive to operate and is not applicable for all elements and transitions. One of the ac Zeeman systems is probably the best choice if cost is not a criterion, based on its applicability to all elements and transitions, analytical performance, and wide number of applications.

4.7 GRAPHITE FURNACE ATOMIC EMISSION SPECTROMETRY

As discussed in Section 2.2, atomic emission involves thermal (nonradiative) promotion of an electron in an atom to an excited state, followed by the release of a photon to a lower energy state (emission) (Fig. 2.1). Atomic emission has been widely used in arcs and sparks, flames (flame photometry), and plasmas (e.g., inductively coupled plasma optical emission spectrometry). Atomic emission can be employed for the determination of nonmetals (S, Cl) not determinable by atomic absorption, and compared to atomic absorption, atomic emission has longer linear calibration graphs (typically 10^3–10^5 compared to 10^2–10^3), and is easier to use in a multielemental mode. Graphite furnace atomic emission spectrometry (GFAES) was developed to employ these potential advantages (83–85). The principal disadvantage of GFAES is the maximum temperature (~2700°C) attainable in a graphite furnace is insufficiently high for efficient thermal excitation of many elements, resulting in relatively poor sensitivity.

To improve the sensitivity, several approaches were developed that employ an auxiliary excitation source to increase the number of excited atoms. One commercially available approach involves the use of a capacitively coupled plasma (CCP) with a conventional graphite furnace for atomic emission, which is called GFCCP-AES, or furnace atomization plasma emission spectrometry (FAPES) (86–88). In the commercial system, a graphite antenna, which is coupled to a 27-MHz radio frequency generator, is inserted into the tube coaxially to establish an argon plasma (Fig. 4.36) (89). A transversely heated tube is incorporated with this system, which should minimize gas-phase interferences.

Table 4.6 shows detection limits obtained by FAPES with a laboratory-constructed instrument for several elements with platform atomization (90). Detection limits are comparable to GFAAS for elements with low excitation energies, but significantly worse for elements with high excitation energies (e.g., arsenic and selenium). Sturgeon and co-workers (91) determined cadmium in

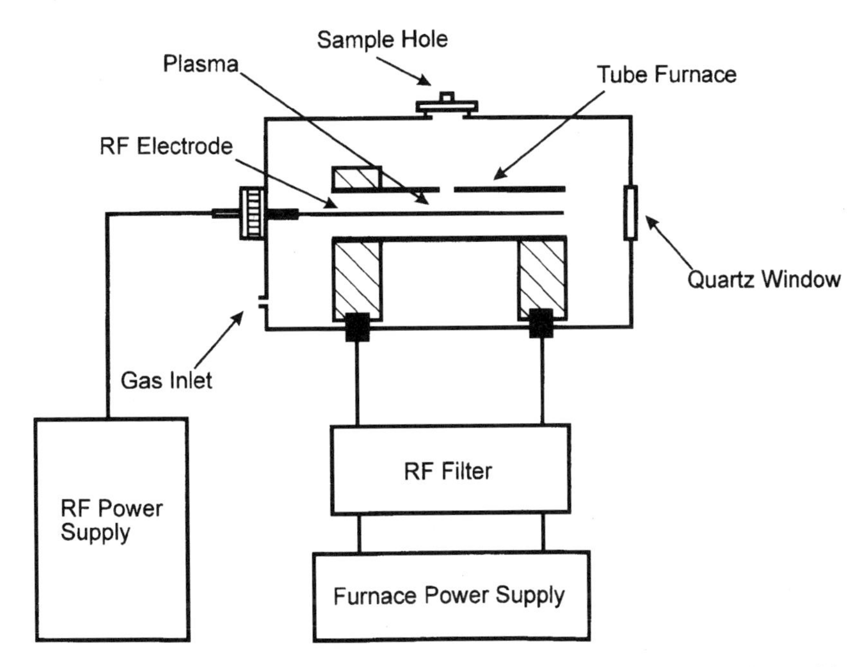

Figure 4.36. Instrumentation for GFCCP-AES. Redrawn with permission from Reference 86.

Table 4.6 Detection Limits of FAPES with Platform Atomization and 0.5 μg Palladium as the Chemical Modifier (90)

Element	Wavelength, nm	Excitation Energy, eV	Atomization Temperature, °C	Detection limit, pg
Ag	328.1	3.78	2100	0.5
As	200.3	7.54	2700	370
Cd	228.8	5.41	2100	2.0
Mn	279.5	4.43	2500	5.0
Pb	283.3	4.37	2100	7.4
Se	196.0	6.32	2700	3600
Sn	284.0	4.79	2500	29
Tl	377.6	3.28	2100	37

sediment and lobster tissue standard reference materials by FAPES. The samples were dissolved using a microwave digestion procedure (Section 6.3.1). Good accuracy was reported compared to the reference values by calibration by the method of standard additions.

The principal advantages of FAPES include the ability to determine nonmetals, which cannot be done by GFAAS (chlorine and sulfur) or are relatively insensitive (phosphorous), and its potential for multielement analysis. However, the RF electrode affects the thermal properties of the furnace, and chemical components may condense on its surface. The development of several commercial multielement GFAAS instruments (Section 4.8.2) allows the use of conventional GFAAS for multielement analysis. In addition, at the present time, there are few applications of GFCCP-AES, and hence a considerable amount of method development would probably be required to use this technique for practical analysis.

4.8 SPECTROMETER DESIGNS

Graphite furnace atomic absorption spectrometers include single-element and multielemental systems. Traditionally GFAAS has been employed as a technique for single-element analysis. This is a significant disadvantage when several elements must be determined in the same sample (92). In addition, single element determinations result in higher costs for consumables, especially for graphite tubes.

Since the late 1980s, several commercial multielement (4–6 elements) GFAAS instruments have been introduced. Multielemental systems clearly allow more rapid analysis, but with the disadvantages of greater complexity, higher cost, and often compromised performance to do simultaneous analysis. In addition, the LDR of GFAAS is generally limited to two to three orders of magnitude. Consequently, it is often not possible to simultaneously determine several elements without dilution due to variations in concentration levels.

4.8.1 Single-Element Instruments

A simple single-beam optical arrangement for GFAAS consists of a hollow cathode lamp, a graphite furnace atomizer, a monochromator, and a detector (Fig. 4.37). The light transmitted in the absence of sample, or baseline, is measured just before the analyte is atomized. The baseline value depends on the lamp, atomizer, and detection system. A high degree of baseline stability is required to obtain the highest precision and accuracy. Early instruments were affected by significant drift in lamp intensity, and hence were frequently operated in a double-beam mode. This arrangement involves separation of the source radiation into two beams, one passing through the atomizer and the other bypassing it. This arrangement allows simultaneous measurement of the reference and transmitted light beams and hence will correct for drift in the

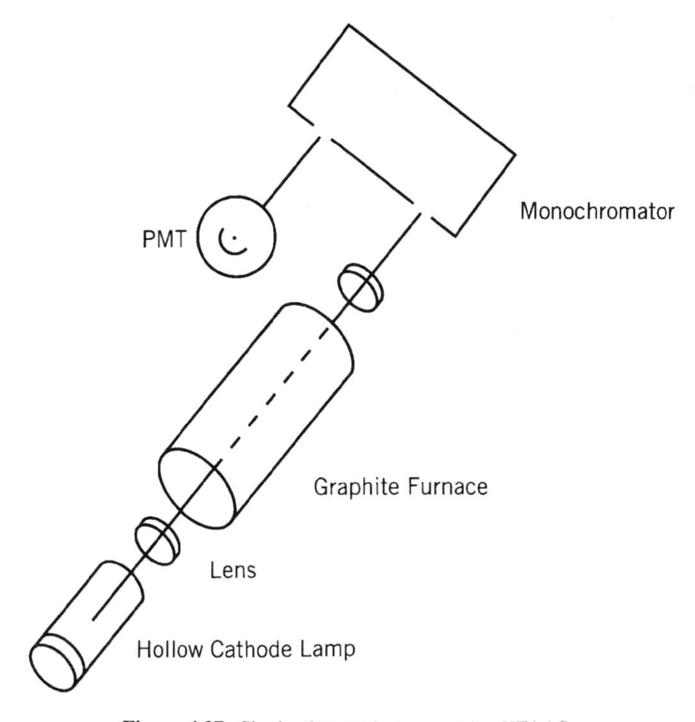

Figure 4.37. Single-element instrument for GFAAS.

source, although it will not correct for drift induced by the atom cell. Most modern instrumentation has sufficiently small lamp drift that a single-beam optical arrangement is employed because of the higher sampling frequency, which results in better precision.

A variety of single-element GFAAS instruments are commercially available equipped with continuum source, self-reversal, or Zeeman background correction (Section 8.1). Several instruments are available with a continuum source and one of the other two methods. The vast majority of GFAAS applications has been performed on single-element instrumentation that has been shown to provide accurate and precise analysis in numerous complex matrices. These instruments are also generally lower in cost than multielement instruments.

4.8.2 Multielement Instruments

Farah and Sneddon (92) recently reviewed instrumentation and applications of multielemental GFAAS. Although a wide variety of approaches have been employed, here we focus on commercial instrumentation and continuum source AAS, which is probably the most thoroughly investigated multielement GFAAS configuration.

Perhaps the most straightforward optical arrangement consists of a separate monochromator and PMT for each element (Fig. 4.38*a*). This design, which has been marketed by Hitachi since the late 1980s, employs four HCLs imaged through a graphite tube with four PMTs, permitting the determination of four elements in a furnace cycle. This commercial system is equipped with dc Zeeman background correction.

Sen Gupta (93,94) determined 17 elements in silicate standard reference materials and rocks with this instrument. The samples were dissolved with hydrofluoric acid and aqua regia, followed by preconcentration by ion-exchange chromatography. The values obtained by GFAAS compared favorably to values obtained by inductively coupled plasma emission and mass spectrometry.

In a second design, marketed by Thermo Jarrell Ash (Franklin, MA), the source beam and light reaching the focal plane of the monochromator are modulated among four source beams and detection wavelengths by use of a mirror and grating controlled by galvanometer (Fig. 4.38*b*). These instruments may employ either continuum source or self-reversal background correction. The sampling frequency of this instrument is relatively low (20 Hz), which is a disadvantage for GFAAS.

Deval and Sneddon (95) used this instrument for the simultaneous determination of lead and cadmium in blood standard reference samples. Acceptable accuracy was achieved by use of self-reversal background correction and an ammonium hydrogen phosphate chemical modifier (Section 5.1).

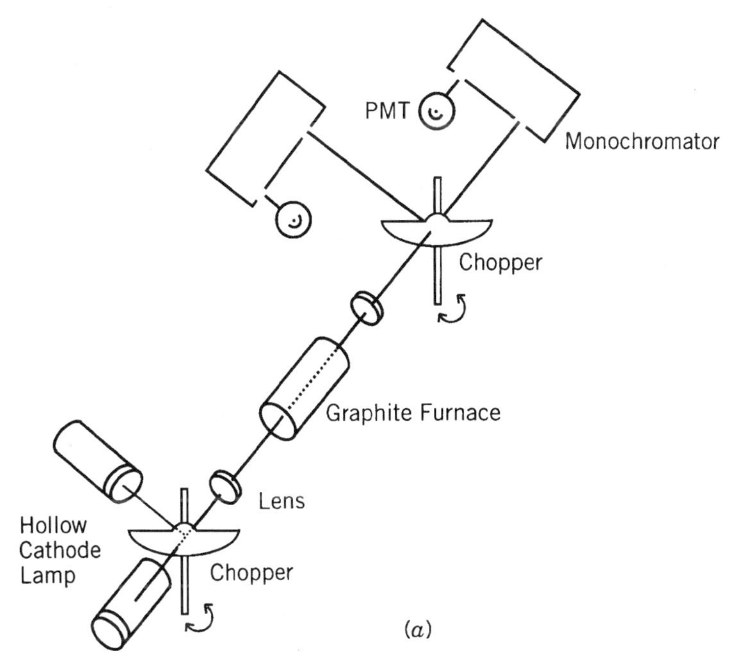

Figure 4.38. Multielement instrument for GFAAS: (*a*) Hitachi design, multiple HCLs with multiple detection systems (note that only two of the four channels present in the Hitachi instrument are illustrated, and the magnet for Zeeman effect background correction has been omitted); (*b*) Thermo Jarrell Ash design, multiple HCLs with one monochromator and PMT with a mirror and grating controlled by a galvanometer; (*c*) Perkin-Elmer design, multiple HCLs with an échelle monochromator and a charge-transfer device; (*d*) Leeman Laboratories design, multiple HCLs arranged on the focal plane of the grating, and (*e*) SIMAAC design, continuum source excitation with a monochromator and PMT.

Sneddon and co-workers (96,97) described the use of this instrument combined with an impaction system for the simultaneous determination of cadmium, chromium, lead, and manganese in aerosols and cigarette smoke (Section 6.2). This system has potential for direct and near real-time (few minutes) of metals in gas samples, but the results were regarded as semiquantitative due to the unavailability of suitable standards.

A third arrangement (Fig. 4.38*c*), introduced by Perkin-Elmer Corporation, employs six HCLs with an échelle monochromator and a solid-state detector (38,59,98). An array of 61 photodiodes was integrated with a complementary metal-oxide-semiconductor (CMOS) charge amplifier array. Diodes are present for the primary resonance line of 39 elements. In addition, 16 secondary lines are also available. The photosensitive area of each detector element is

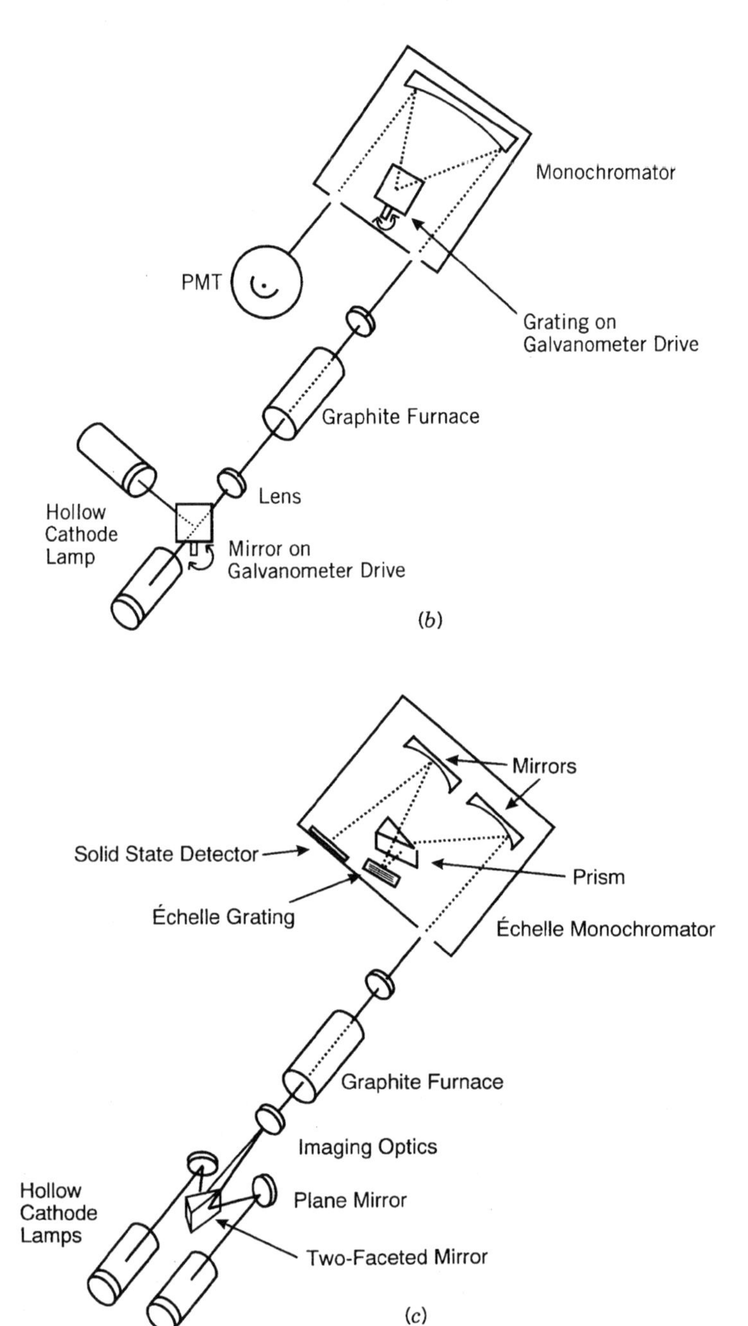

Figure 4.38. *(continued)*

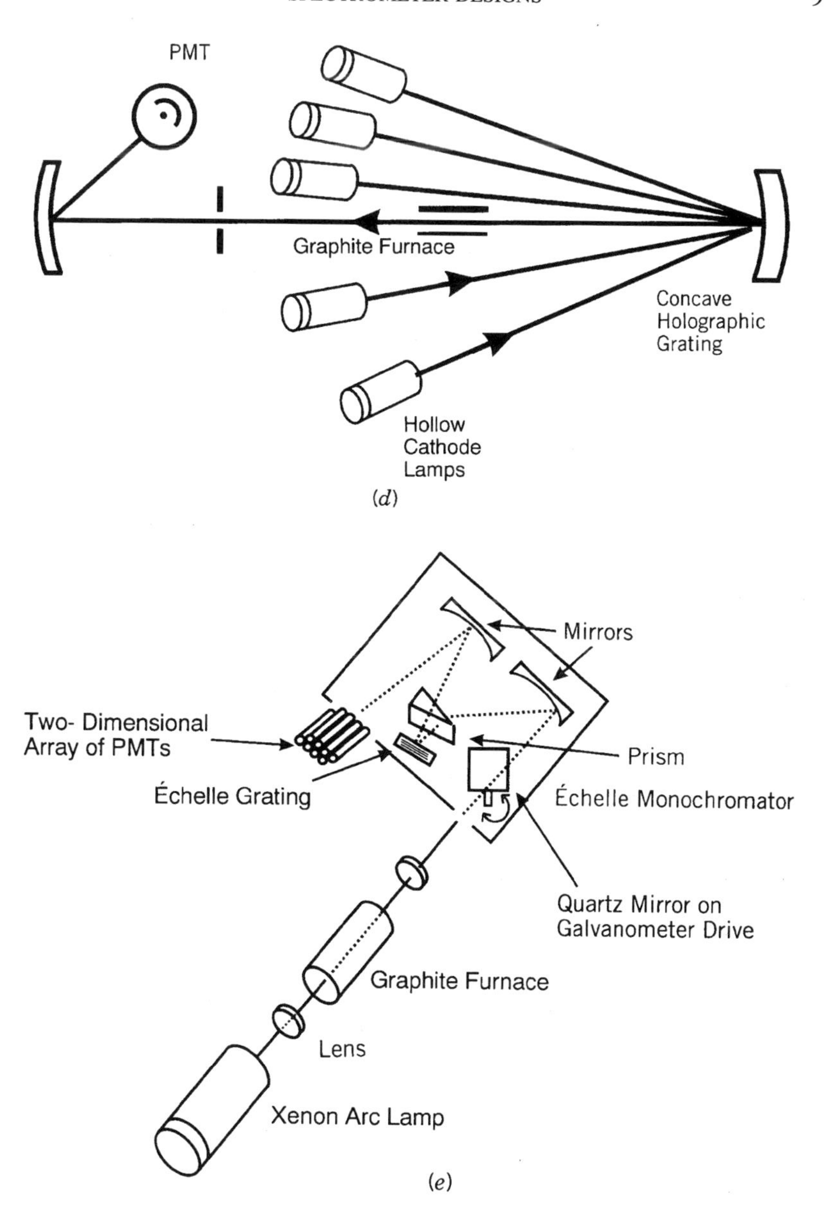

Figure 4.38. *(continued)*

approximately 1 by 2 mm. A transversely heated furnace was employed with inverse, longitudinal ac Zeeman effect background correction.

Harnly and Radziuk (38) evaluated the ability of this instrument to simultaneously determine cadmium, copper, chromium, lead, and vanadium. Cadmium and lead are volatile elements that may be lost at relatively low pyrolysis temperatures (<500°C without a chemical modifier) and atomize below 2000°C, while vanadium is a refractory element. Normally one would expect that a compromise in pyrolysis and atomization temperatures must be employed for this set of elements. However, these authors reported that the most suitable atomization temperature for the simultaneous determination of these elements was 2500°C because the best signal-to-noise ratio for each element was obtained at this temperature. In addition, end-capped THGA tubes provided enhanced sensitivities compared to standard THGA tubes.

A fourth arrangement, available commercially from Leeman Laboratories (Lowell, MA), has up to six HCLs oriented along the focal plane of the grating, with the lamp emission focused onto the grating and directed through the graphite furnace (Fig. 4.38*d*) (99). An off-axis mirror focuses the transmitted beams onto a photomultiplier tube. Background correction is achieved by use of the self-reversal method. This instrument was originally designed for the determination of arsenic, lead, selenium, antimony, and thallium, although other elements may also be determined with the system. However, all elements must be specified when the instrument is purchased, and hence this instrument is only suitable for users who determine a specific group of six or fewer elements. The accuracy of this instrument was evaluated by the determination of five elements in a water standard reference material. Good accuracy was reported for this analysis.

Another major type of multielemental GFAAS instrument employs a xenon arc lamp (continuum source) with a high-resolution monochromator to provide significantly high spectral resolution (9,10) (Fig. 4.38*e*). This instrument design was named simultaneous multielement atomic absorption with a continuum source (SIMAAC). One of the earlier designs employed an échelle polychromator with up to 16 PMTs, one for each analyte. Its principal advantage for practical analysis is an increased linear dynamic range of up to five orders of magnitude, compared to two orders of magnitude with a line source, by scanning the monochromator away from the center of the analytical wavelength at relatively high concentrations. However, the sensitivity was a factor of 2 worse than line source AAS for most elements and a factor of 10 worse for elements whose analytical wavelengths are below 250 nm (e.g., Cd, Zn). More recently, a photodiode array replaced the PMTs and gave comparable sensitivity to line source instruments (8). A third configuration employed a 1-m focal length monochromator and a linear photodiode array that could monitor a 10-nm spectral region at a given time (11).

Harnly and colleagues (8–10,100–103) have done a number of studies to evaluate the applicability of the SIMAAC system to do practical analysis. Harnly and Kane (100) simultaneously determined nine elements in a National Institute of Standards and Technology acidified water standard reference material. Accuracies and precisions of ±10 to 15% were reported. Harnly and co-workers also evaluated this instrument for the determination of several elements in biological standard reference materials by platform (104) and probe (101) atomization. In general, the accuracy and precision of the platform results were superior to the probe data.

REFERENCES

1. A. Walsh, *Spectrochim. Acta*, **7**, 108 (1955).
2. D. J. Butcher, J. P. Dougherty, J. T. McCaffrey, F. R. Preli, A. P. Walton, and R. G. Michel, *Prog. Anal. Spectrosc.*, **10**, 359 (1987).
3. S. J. Haswell, "Instrumental Requirements and Optimization," in S. J. Haswell, Ed., *Atomic Absorption Spectrometry; Theory, Design, and Applications*, Elsevier, Amsterdam, 1991.
4. L. H. J. Lajunen, *Spectrochemical Analysis by Atomic Absorption and Emission*, Royal Society of Chemistry, Cambridge, England, 1992.
5. B. Welz, *Atomic Absorption Spectrometry*, 2nd ed., VCH, Weinheim, Germany, 1985.
6. F. Paschen, *Ann. Phys.*, **50**, 901 (1916).
7. J. Sneddon, R. F. Browner, P. N. Keliher, J. D. Winefordner, D. J. Butcher, and R. G. Michel, *Prog. Anal. Spectrosc.*, **12**, 369 (1989).
8. J. M. Harnly, *J. Anal. Atom. Spectrom.*, **8**, 317 (1993).
9. J. M. Harnly, T. C. O'Haver, B. Golden, and W. R. Wolf, *Anal. Chem.*, **51**, 2007 (1979).
10. J. M. Harnly and D. L. Garland, "Multielement Atomic Absorption Methods of Analysis," in J. F. Riordan and B. L. Vallee, Eds., *Metallobiochemistry, Part A*, Academic, San Diego, 1988.
11. R. Fernando and B. T. Jones, *Spectrochim. Acta, Part B*, **49B**, 615 (1994).
12. K. Niemax, A. Zybin, C. Schnürer-Patschan, and H. Groll, *Anal. Chem.*, **68**, 351A (1996).
13. K. Niemax, H. Groll, and C. Schnürer-Patschan, *Spectrochim. Acta Rev.*, **15**, 349 (1993).
14. H. Groll and K. Niemax, *Spectrochim. Acta, Part B*, **48B**, 633 (1993).
15. C. Schnürer-Patschan, A. Zybin, H. Groll, and K. Niemax, *J. Anal. Atom. Spectrom.*, **8**, 1103 (1993).
16. J. Sneddon, L. Bezur, R. G. Michel, and J. M. Ottaway, *Anal. Proc.*, **19**, 35 (1982).

17. J. D. Ingle and S. R. Crouch, *Spectrochemical Analysis*, Prentice-Hall, Englewood Cliffs, NJ, 1988.
18. B. V. L'vov, *Ing. Fiz. Zh.*, **2**, 44 (1959).
19. B. V. L'vov, *Ing. Fiz. Zh.*, **2**, 56 (1959).
20. B. V. L'vov, *Spectrochim. Acta*, **17**, 761 (1961).
21. B. V. L'vov, *Spectrochim. Acta, Part B*, **39B**, 149 (1984).
22. B. V. L'vov, *J. Anal. Atom. Spectrom.*, **3**, 9 (1988).
23. B. V. L'vov, *Anal. Chem.*, **63**, 924A (1991).
24. H. Falk, *CRC Crit. Rev. Anal. Chem.*, **19**, 29 (1988).
25. J. W. Robinson, *Anal. Chem.*, **66**, 472A (1994).
26. S. J. Haswell, Ed. *Atomic Absorption Spectrometry; Theory, Design, and Applications*, Analytical Spectroscopy Library Ed., Vol. 5, Elsevier, Amsterdam, 1991.
27. A. Walsh, *Anal. Chem.*, **63**, 933A (1991).
28. H. Massmann, *Spectrochim. Acta, Part B*, **23B**, 215 (1968).
29. B. Welz, G. Schlemmer, H. M. Ortner, and W. Wegscheider, *Prog. Anal. Spectrosc.*, **12**, 111 (1989).
30. E. Bulska and W. Jedral, *J. Anal. Atom. Spectrom.*, **10**, 49 (1995).
31. G. Müller-Vogt, F. Weigend, and W. Wendl, *Spectrochim. Acta, Part B*, **51B**, 1133 (1996).
32. M. A. Alvarez, N. Carrión, and H. Gutiérrez, *Spectrochim. Acta, Part B*, **51B**, 1121 (1996).
33. B. Welz, M. Sperling, G. Schlemmer, N. Wenzel, and G. Marowsky, *Spectrochim. Acta, Part B*, **43B**, 1187 (1988).
34. W. Frech, D. C. Baxter, and B. Hütsch, *Anal. Chem.*, **58**, 1973 (1986).
35. M. Sperling, B. Welz, J. Hertzberg, C. Rieck, and G. Marowsky, *Spectrochim. Acta, Part B*, **51B**, 897 (1996).
36. W. Frech, B. V. L'vov, and N. P. Romanova, *Spectrochim. Acta, Part B*, **47B**, 1461 (1992).
37. W. Frech and B. V. L'vov, *Spectrochim. Acta, Part B*, **48B**, 1371 (1993).
38. J. M. Harnly and B. Radziuk, *J. Anal. Atom. Spectrom.*, **10**, 197 (1995).
39. P. R. Boulo, J. J. Soraghan, D. A. Sadler, D. Littlejohn, and A. Creeke, *J. Anal. Atom. Spectrom.*, **12**, 293 (1997).
40. G. Donati, M. Ottaviani, and E. Veschetti, *Microchem. J.*, **54**, 287 (1996).
41. J. J. Sotera, L. C. Cristiano, M. K. Conley, and H. L. Kahn, *Anal. Chem.*, **55**, 204 (1983).
42. J. Versieck and L. Vanballenberghe, "Collection, Transport, and Storage of Biological Samples for the Determination of Trace Metals," in H. G. Seiler, A. Sigel, and H. Sigel, Eds. *Handbook on Metals in Clinical and Analytical Chemistry*, Marcel Dekker, New York, 1994.
43. I. L. Shuttler and H. T. Delves, *Analyst*, **111**, 651 (1986).

44. R. H. Eckerlin, D. W. Hoult, and G. R. Carnrick, *Atom. Spectrosc.*, **8**, 64 (1987).
45. M. Hoenig and A. Cilissen, *Spectrochim. Acta, Part B*, **48B**, 1003 (1993).
46. P. J. Parsons and W. Slavin, *Spectrochim. Acta, Part B*, **48B**, 925 (1993).
47. Z. Li, G. Carnrick, and W. Slavin, *Spectrochim. Acta, Part B*, **48B**, 1435 (1993).
48. L. Lian, *Spectrochim. Acta, Part B*, **47B**, 239 (1992).
49. D. J. Halls, *J. Anal. Atom. Spectrom.*, **10**, 169 (1995).
50. B. V. L'vov, *Spectrochim. Acta, Part B*, **33B**, 153 (1978).
51. O. O. Ajayi, D. Littlejohn, and C. B. Boss, *Talanta*, **36**, 805 (1989).
52. D. Littlejohn and J. M. Ottaway, *Analyst*, **104**, 1138 (1979).
53. S. P. Corr and D. Littlejohn, *J. Anal. Atom. Spectrom.*, **3**, 125 (1988).
54. W. Frech and S. Jonsson, *Spectrochim. Acta, Part B*, **37B**, 1021 (1982).
55. E. Lundberg, W. Frech, D. C. Baxter, and A. Cedergren, *Spectrochim. Acta, Part B*, **43B**, 451 (1988).
56. D. A. Skoog and J. J. Leary, *Principles of Instrumental Analysis*, 4th ed., Saunders, Fort Worth, TX, 1992.
57. D. A. Skoog, D. M. West, and F. J. Holler, *Fundamentals of Analytical Chemistry*, 7th ed., Saunders College Publishing, Fort Worth, TX, 1996.
58. H. A. Stroble and W. R. Heineman, *Chemical Instrumentation: A Systematic Approach*, 3rd ed., Wiley, New York, 1989.
59. B. Radziuk, G. Rödel, H. Stenz, H. Becker-Ross, and S. Florek, *J. Anal. Atom. Spectrom.*, **10** 127 (1995).
60. J. V. Sweedler, R. B. Bilhorn, P. M. Epperson, G. R. Sims, and M. B. Denton, *Anal. Chem.*, **60**, 282A (1988).
61. J. V. Sweedler, R. B. Bilhorn, P. M. Epperson, G. R. Sims, and M. B. Denton, *Anal. Chem.*, **60**, 327A (1988).
62. Q. S. Hanley, C. W. Earle, F. M. Pennebaker, S. P. Madden, and M. B. Denton, *Anal. Chem.*, **68**, 661A (1996).
63. J. McNally and J. A. Holcombe, *Anal. Chem.*, **63**, 1918 (1991).
64. R. A. Newstead, W. J. Price, and P. J. Whiteside, *Prog. Anal. Atom. Spectrosc.*, **1**, 267 (1978).
65. J. Sneddon, *Spectroscopy*, **2**(5), 38 (1986).
66. W. Slavin and G. R. Carnrick, *CRC Crit. Rev. Anal. Chem.*, **19**, 95 (1988).
67. G. Dulude, "Background Correction Techniques in Atomic Absorption Spectrometry," in J. Sneddon, Ed., *Advances in Atomic Spectroscopy*, Jai Press, Greenwich, CT, 1992.
68. C. L. Sanford, S. E. Thomas, and B. T. Jones, *Appl. Spectrosc.*, **50**, 174 (1996).
69. S. R. Koirtyohann and E. E. Pickett, *Anal. Chem.*, **37**, 601 (1965).
70. S. B. Smith and G. M. Hieftje, *Appl. Spectrosc.*, **37**, 419 (1983).
71. M. T. C. de Loos-Vollebregt and L. de Galan, *Prog. Anal. Atom. Spectrosc.*, **8**, 47 (1985).

72. K. Yasuda, H. Koizumi, K. Ohishi, and T. Noda, *Prog. Anal. Atom. Spectrosc.*, **3**, 299 (1980).
73. M. T. C. de Loos-Vollebregt, L. de Galan, and J. W. M. van Uffelen, *Spectrochim. Acta, Part B*, **43B**, 1147 (1988).
74. M. Berglund, W. Frech, and D. C. Baxter, *Spectrochim. Acta, Part B*, **46B**, 1767 (1991).
75. Y. Y. Zong, P. J. Parsons, and W. Slavin, *Spectrochim. Acta, Part B*, **49B**, 1667 (1994).
76. U. Heitmann, M. Schütz, H. Becker-Roß, and S. Florek, *Spectrochim. Acta, Part B*, **51**, 1095 (1996).
77. B. V. L'vov, L. K. Polzik, N. V. Kocharova, Y. A. Nemets, and A. V. Novichikhin, *Spectrochim. Acta, Part B*, **47B**, 1187 (1992).
78. E. G. Su, A. I. Yuzefovsky, R. G. Michel, J. T. McCaffrey, and W. Slavin, *Spectrochim. Acta, Part B*, **49B**, 367 (1994).
79. A. I. Yuzefovsky, R. F. Lonardo, J. X. Zhou, R. G. Michel, and I. Koltracht, *Spectrochim. Acta, Part B*, **51B**, 713 (1996).
80. R. F. Lonardo, A. I. Yuzefovsky, J. X. Zhou, J. T. McCaffrey, and R. G. Michel, *Spectrochim. Acta, Part B*, **51B**, 1309 (1996).
81. A. I. Yuzefovsky, R. F. Lonardo, J. X. Zhou, and R. G. Michel, *Appl. Spectrosc.*, **51**, 738 (1997).
82. M. T. C. de Loos-Vollebregt and L. de Galan, *Spectrochim. Acta, Part B*, **35B**, 495 (1980).
83. L. Bezur, J. Marshall, J. M. Ottaway, and R. Fakhrul-Aldeen, *Analyst*, **108**, 553 (1983).
84. D. C. Baxter and W. Frech, *Fresenius J. Anal. Chem.*, **328**, 324 (1987).
85. D. C. Baxter and W. Frech, *Spectrochim. Acta, Part B*, **50B**, 655 (1995).
86. D. C. Liang and M. W. Blades, *Spectrochim. Acta*, **44B**, 1059 (1989).
87. R. E. Sturgeon, S. N. Willie, V. Luong, S. S. Berman, and J. G. Dunn, *J. Anal. Atom. Spectrom.*, **4**, 669 (1989).
88. R. E. Sturgeon, S. N. Willie, V. Luong, and S. S. Berman, *Anal. Chem.*, **62**, 2370 (1990).
89. P. R. Banks, D. C. Liang, and M. W. Blades, *Spectroscopy*, **7**(8), 36 (1992).
90. R. E. Sturgeon, S. N. Willie, V. T. Luong, and S. S. Berman, *J. Anal. Atom. Spectrom.*, **6**, 19 (1991).
91. R. E. Sturgeon, S. N. Willie, V. T. Luong, and S. S. Berman, *J. Anal. Atom. Spectrom.*, **5**, 635 (1990).
92. K. S. Farah and J. Sneddon, *Appl. Spectrosc. Rev.*, **30**, 351 (1995).
93. J. G. Sen Gupta, *Talanta*, **40**, 791 (1993).
94. J. G. Sen Gupta, *J. Anal. Atom. Spectrom.*, **8**, 93 (1993).
95. A. Deval and J. Sneddon, *Microchem. J.*, **52**, 96 (1995).
96. J. Sneddon, M. V. Smith, S. Indurthy, and Y.-I. Lee, *Spectroscopy*, **10**(1), 26 (1995).

97. Y.-I. Lee, M. V. Smith, S. Indurthy, A. Deval, and J. Sneddon, *Spectrochim. Acta, Part B*, **51B**, 109 (1996).
98. B. Radziuk, G. Rödel, M. Zeiher, S. Mizuno, and K. Yamamoto, *J. Anal. Atom. Spectrom.*, **10**, 415 (1995).
99. D. R. Demers and M. C. Almeida, *Am. Environ. Lab.*, **June**, 13 (1995).
100. J. M. Harnly and J. S. Kane, *Anal. Chem.*, **56**, 48 (1984).
101. J. Carroll, N. J. Miller-Ihli, J. M. Harnly, D. Littlejohn, J. M. Ottaway, and T. C. O'Haver, *Analyst*, **110**, 1153 (1985).
102. C. M. M. Smith and J. M. Harnly, *Spectrochim. Acta, Part B*, **49B**, 387 (1994).
103. C. M. Smith and J. M. Harnly, *J. Anal. Atom. Spectrom.*, **10**, 187 (1995).
104. J. M. Harnly, N. J. Miller-Ihli, and T. C. O'Haver, *Spectrochim. Acta, Part B*, **39B**, 305 (1984).
105. W. Slavin and G. R. Carnrick, *Atom. Spectrosc.*, **7**, 9 (1986).
106. R. Wennrich, W. Frech, and E. Lundberg, *Spectrochim. Acta, Part B*, **44B**, 239 (1989).
107. K. E. A. Ohlsson and W. Frech, *J. Anal. Atom. Spectrom.*, **4**, 379 (1989).
108. G. Wibetoe and F. J. Langmyhr, *Anal. Chim. Acta*, **198**, 81 (1987).
109. P. S. Doidge, *Spectrochim. Acta, Part B*, **46B**, 1779 (1991).
110. G. Wibetoe and F. J. Langmyhr, *Anal. Chim. Acta*, **186**, 155 (1986).
111. U. Kurfürst and J. Pauwels, *J. Anal. Atom. Spectrom.*, **9**, 531 (1994).

CHAPTER 5

INTERFERENCE-FREE ANALYSIS

A potential problem for accurate elemental analysis by GFAAS is a change, either positive or negative, in the analyte absorbance of samples compared to standard solutions, resulting in inaccurate analyses. These interferences, which are characterized as spectral, chemical, or physical, are often referred to as *matrix effects*. One approach to deal with matrix effects is to match the composition of the standard solutions to the sample solutions, which is called *matrix matching*. Matrix matching is inconvenient and often impractical because of the impossibility of matching all components in samples. An alternative calibration technique to alleviate some matrix effects is the method of standard additions (Section 3.3). During the 1980s and 1990s, modern furnace technology (Table 1.1) was developed, reducing the severity of matrix effects and hence the need for matrix matching and the method of standard additions. This chapter discusses types of interferences encountered in GFAAS and methods to eliminate them.

5.1 SPECTRAL INTERFERENCES

Spectral interferences are caused by the presence of concomitants that affect the quantity of source radiation that reaches the detection system. They are additive interferences that increase or decrease all measurements equally (Fig. 5.1). There are two general types of spectral interferences: the presence of a nonanalyte atomic absorption line close to the absorption wavelength of the analyte (spectral overlaps) and nonspecific reduction of the source intensity (background attenuation), which is induced by scatter of source radiation due to particles, or absorption of light by molecules, from the sample matrix.

5.1.1 Spectral Overlaps

Spectral overlaps (1,2) in atomic absorption involve the presence of a nonanalyte element that absorbs source radiation. These overlaps are generally rare because there are fewer absorption lines than emission lines, and the spectral output of an

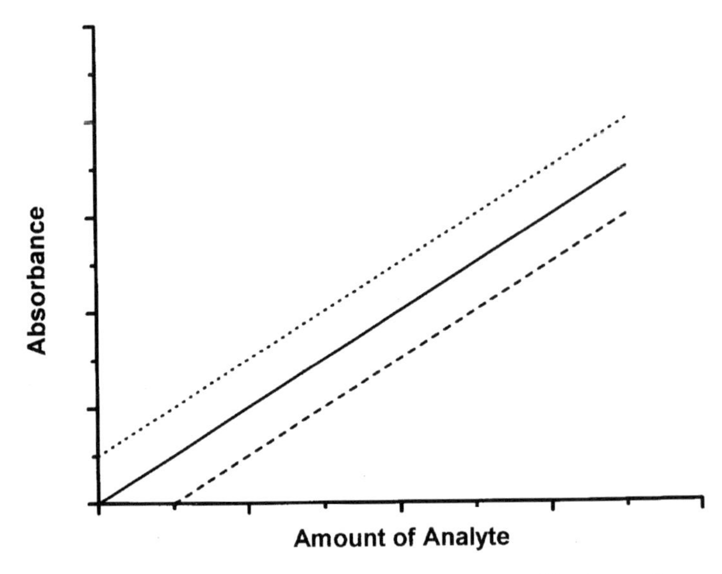

Figure 5.1. Effects of additive (e.g., spectral) interferences upon a calibration graph: —, no additive interference; · · ·, positive additive interference; - - -, negative additive interference.

HCL is relatively narrow compared to the absorption profile (Section 4.1.1). In practice, overlaps only occur in the relatively rare case when absorption lines of two elements are within 0.01 nm of each other. In addition, the interfering element must usually be present in relatively high concentrations in order to cause a significant interference.

Some examples of interferences caused by spectral overlaps are given in Table 5.1. The interference with the greatest analytical significance occurs between gallium and manganese at 403.3 nm (1). In general, spectral overlaps can be avoided by the use of an alternative wavelength. The use of multielement

Table 5.1 Examples of Spectral Interferences Caused by Spectral Overlaps (1)

Analyte; Emission line, nm	Interferent; Absorption line, nm	Intensity Ratio
Al; 308.215	V; 308.211	200 : 1
Cu; 324.754	Eu; 324.753	500 : 1
Fe; 271.903	Pt; 271.904	500 : 1
Ga; 403.298	Mn; 403.307	3 : 1
Hg; 253.652	Co; 253.649	8 : 1
Sb; 217.023	Pb; 216.999	10 : 1
Si; 250.690	V; 250.690	8 : 1

HCLs may increase the risk of spectral interferences if, in addition to the analytical line, a spectral line of a concomitant element is within the monochromator bandpass. In this situation, the absorbance depends on the concentration of both elements, and hence specificity is lost.

It should be pointed out that nonanalyte emission is not an interference in atomic absorption because modulation of the source allows discrimination against dc signals, such as emission, as discussed in Section 4.1.

5.1.2 Background Attenuation

Background attenuation involves scattering of light by particles or absorption of light by molecules, which reduces the intensity of the source beam, and increases the absorbance, giving inaccurately high concentration measurements. Particles that induce *scatter* may be formed by the decomposition of organic material or by condensation of inorganic salts at relatively cool regions at the ends of the furnace. The magnitude of the scattering coefficient τ is given by Rayleigh's law of scattering:

$$\tau = 24\pi^3 \frac{Nv^2}{\lambda^4} \tag{5.1}$$

where N is the density of scatters (number/cm^3), v is the volume of the scatters (cm^3), and λ is the wavelength of light (cm). This relationship shows that scatter has a maximum value with large scattering particles and a short wavelength.

Scatter may be induced by the condensation of volatile concomitants onto relatively cool regions of a graphite tube. The vaporization of organic materials may produce smoke that serves as a scattering medium. A third source of scatter is sublimation of graphite, although this problem has been minimized with modern tube furnace designs.

Molecular absorption involves the absorption of light by small molecules from the sample matrix. In general, halide compounds absorb significant amounts of light between 200 and 400 nm, which can be characterized as photodissociation continuum or electronic band spectra (3–5). The former are characterized by broad spectra induced by the absorption of light that causes dissociation of the molecule (Fig. 5.2). Molecular electronic band spectra are typically 0.5 to 10 nm in width and composed of several vibrational lines that are further split into rotational lines (6). The result is that the background is generally not constant across the width of molecular band (e.g., Fig. 4.31). A list of the absorption wavelengths of some diatomic molecules is given in Table 5.2. As discussed in Section 4.6.2, continuum source background correction would not be expected to give accurate results in the presence of a molecular band because of the variation in the background level across the spectral bandpass of

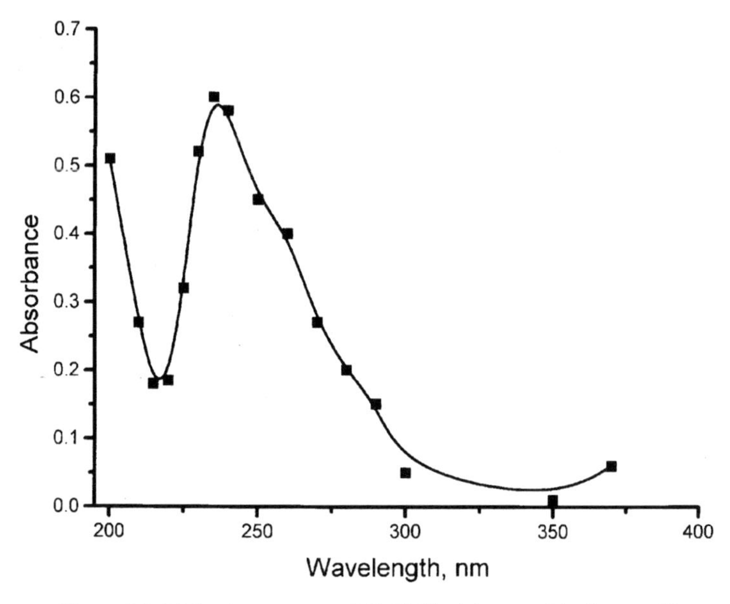

Figure 5.2. Molecular spectrum of NaCl (5 μg) in a graphite furnace (3).

the monochromator. Self-reversal or Zeeman effect background correction should be employed in these cases, although these methods cannot accurately correct for all interferences (Fig. 4.31).

Vadja (7) discussed atomic lines that are sufficiently close to resonance lines of other elements and potentially could induce background correction errors using a continuum source (Table 4.3). For example, Manning (8) showed that

Table 5.2 Spectral Band Interferences (6)

Molecule	Wavelength Range, nm
AlF	227.4–227.8
GaF	211–213
InF	233.3–234.2
AlCl	259–264
GaCl	246–251
InCl	246–251
AlBr	274–284
GaBr	262–272
InBr	280–290
CaF	603–607
CaCl	617–622

high concentrations of iron induce continuum source background interferences for selenium determination at 196.0 nm. The use of an alternative selenium wavelength (204.0 nm) or a modern correction method eliminates these interferences. Some background correction errors have been reported with self-reversal and Zeeman background correction, as discussed in Sections 4.6.3 and 4.6.4 and Table 4.4 (2,9,10).

The source of background signals, either scatter or molecular absorption, is often difficult to determine. Massmann et al. (11,12) reported that the precision of scatter signals was worse than for molecular absorption. In addition, scatter was observed later in the furnace cycle than molecular absorption. The latter effect was attributed to migration of molecules to the cooler ends of the furnace where their condensation induced scattering. In general, the spectral characteristics of both types of background attenuation are sufficiently similar that they can be corrected with the same methods, and hence for practical analysis it is not necessary to identify the type of background signal present.

Kurfürst and Pauwels (13) reported a spectral interference for lead at the 283.2-nm line in coal determined by solid sampling that was attributed to S_2 molecules. The interference was removed by dissolution of the sample. Cabon and Bihan (14) observed a spectral interference for the determination of zinc (213.9) induced by Zeeman splitting of NO absorption bands. The NO was produced by the decomposition of nitric acid that was employed as a chemical modifier (discussed further in this section). The interference was removed by the use of a pyrolysis temperature of 850°C. Several authors (1,2,13) have summarized spectral interferences reported in the literature.

Background can be minimized by several protocols included in modern furnace technology. First, as discussed in Section 4.6, background correction methods can be used to measure the background and subtract it from the uncorrected measurement to give a background corrected signal. The self-reversal and Zeeman effect methods have the ability to correct for structured background, which continuum source background correction cannot (Section 4.6).

Second, it is important to use the combination of a furnace with fast heating rates with platform or probe atomization. This system helps ensure that atomization occurs into a relatively hot environment that reduces molecule and particle formation. If available, the use of a transversely heated furnace may help minimize spectral interferences. This design has reduced spatial temperature gradients compared to a longitudinally heated furnace (15) and therefore causes a decrease in background. A further reduction in background can be achieved by the use of end caps, which reduce gas-phase temperature gradients at the end of the tube.

A third technique to reduce background is the addition of chemical reagents, called *chemical* or *matrix modifiers* (1,16–19). For volatile elements, a chemical modifier serves to reduce the volatility of the analyte or to increase the volatility of the matrix. In general, it is difficult to ensure that all matrix species are removed, and hence most modifiers have served to decrease volatility of the analyte. Chemical modifiers allow the use of a higher pyrolysis temperature to vaporize more matrix components during the pyrolysis step, and hence minimize scatter and molecule formation during the atomization step when the analytical measurement is performed. An example of the use of modifiers (palladium alone or with other modifiers) to stabilize an analyte (lead) is demonstrated in Figure 5.3.

Modifiers may introduce some problems for real sample analysis. First, they may require the use of a higher atomization temperature, which may reduce the characteristic mass due to a higher rate of diffusion from the tube. Figure 5.4 shows the broadening of the absorbance signal as higher concentrations of palladium are used for the determination of lead with a constant atomization temperature. Second, Frech and L'vov (20) showed that analyte may condense with a chemical modifier in cool regions of a graphite tube, which is called *matrix trapping*. Third, their presence may produce spectral interferences. Zong et al. (9) and Heitmann et al. (10) (Fig. 4.31) reported Zeeman effect background correction errors for the determination of alkaline earth metals due to PO molecules generated from an ammonium dihydrogen phosphate modifier.

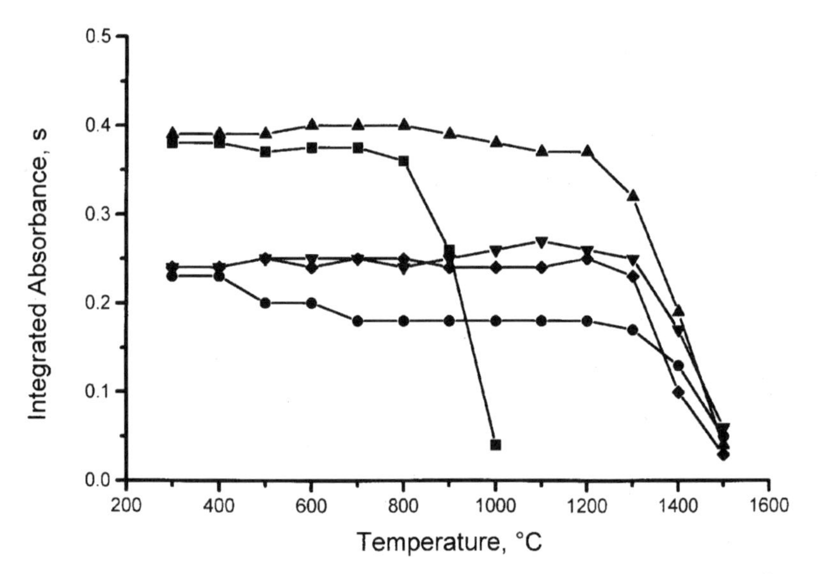

Figure 5.3. Pyrolysis optimization graph for 1-ng lead (■) alone; with palladium 8 μg (●); with palladium 8 μg and magnesium 6 μg (▲) with palladium 8 μg and ascorbic acid 60 μg (▼); with palladium 8 μg and carbon 24.5 μg (◆). Atomization temperature: 1900°C; wavelength: 283.3 nm (43).

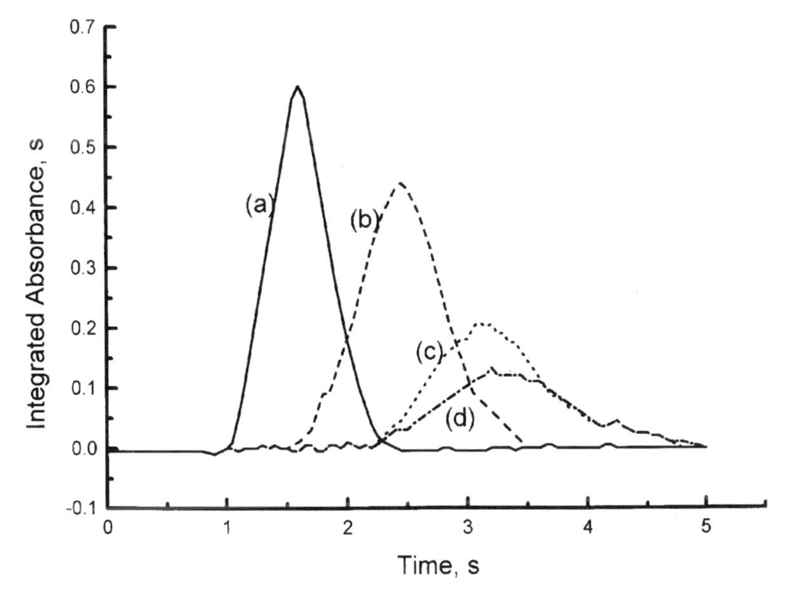

Figure 5.4. Effect of palladium (as the nitrate) upon the absorbance profile of 1.0 ng of lead: (*a*) no palladium added; (*b*) 1.0 μg palladium; (*c*) 5.0 μg palladium; (*d*) 15 μg palladium. Atomization temperature: 1900 °C; pyrolysis temperature; 900 °C; wavelength: 283.3 nm (44).

A chemical modifier must be available in high purity, must stabilize the analyte to a high pyrolysis temperature, cannot significantly reduce the lifetime of the graphite furnace, and cannot cause spectral interferences. In addition, an ideal modifier would be applicable to a wide variety of elements and not contain an element in high concentration that may be determined at trace levels. Tsalev (16) has summarized the various substances that have been used as chemical modifiers. Commonly used modifiers include magnesium nitrate, ammonium dihydrogen or diammonium hydrogen phosphate, nickel (nitrate), and palladium (nitrate).

Palladium has been widely studied in recent years because it best meets the criteria described above: availability in high purity, no effect upon tube lifetime, no production of interferences, and is rarely determined by GFAAS. Moreover, palladium, either alone or with another modifier, such as magnesium nitrate, has been used for a variety of elements including antimony, arsenic, bismuth, cadmium, chromium, copper, indium, lead, manganese, selenium, tellurium, thallium, tin, and vanadium (21–24).

Schlemmer and Welz (21) investigated the use of palladium and magnesium nitrates for the determination of nine elements. Table 5.3 shows that a higher pyrolysis temperature could be used for most elements compared to the cases of

Table 5.3 Maximum Pyrolysis Temperature without Loss of Analyte (21)

	Maximum Pyrolysis Temperature, °C		
Element	No Modifier	Other Modifier	$Pd(NO_3)_2 + Mg(NO_3)_2$
As	200	Ni^{2+}, 1300	1200
Bi	600	Ni^{2+}, 900	1200
In	700	—	1200
Pb	500	$PO_4^{3-} + Mg^{2+}$, 900	1000
Sb	900	Ni^{2+}, 1100	1300
Se	200	$Ni^{2+} + Mg^{2+}$, 900	1100
Sn	800	$PO_4^{3-} + Mg^{2+}$, 800	1400
Te	700	Ni^{2+}, 1000	1200
Tl	600	H_2SO_4, 600	900

no modifier or with other commonly used modifiers. They demonstrated good accuracy for the determination of selenium in blood serum with palladium, as compared to lower, inaccurate results with other modifiers.

Harnly and Radziuk (22) investigated the simultaneous determination of cadmium, chromium, copper, lead, and vanadium using a transversely heated graphite furnace with and without a palladium nitrate/magnesium nitrate chemical modifier. The most suitable atomization temperature for the determination of these elements was 2500°C. Although matrix trapping was observed using the modifier without end caps, this interference was removed using an end-capped furnace. The modifier allowed the use of a pyrolysis temperature of 700°C, compared to 400°C in its absence.

The use of oxygen as a chemical modifier is called *oxygen ashing* (25–30). Oxygen or air is introduced into the tube during the pyrolysis step at temperatures up to 800°C. This procedure has been most commonly employed for the analysis of biological samples because it prevents formation of carbonaceous residue produced with conventional pyrolysis. The residue may prevent reproducible sample introduction, affect atomization rates, and block the source beam. Together these effects may cause a degradation in sensitivity and precision. Oxygen ashing has been shown to prevent formation of this residue and also has been successful at reducing background in these matrices.

Shuttler and Delves (25) determined lead in whole blood by GFAAS with oxygen ashing at 550°C and phosphate modifiers. The sample preparation involved simple dilution of the sample. Good accuracy and precision were reported using this method. Eckerlin et al. (29) reported that oxygen ashing did not completely prevent the formation of carbonaceous residue, and that it was necessary to occasionally blow compressed air through the dosing hole with a furnace window off to remove it.

Although the majority of research on modifiers has focused on the reduction of the volatility of volatile elements, they have also been employed to increase the volatility of refractory and carbide-forming elements (31–33). Nater et al. (31) employed hydrofluoric acid and cesium fluoride as modifiers for the determination of aluminum. This method was reported to induce vaporization of aluminum as aluminum trifluoride, which rapidly decomposed to produce sharp atomic absorption peaks. Metals did not interfere with this protocol, but organic compounds such as ethylenediaminetetraacetic acid (EDTA) and phthalic acid caused a suppression of the aluminum signal. Ericson and co-workers reported that barium fluoride had similar positive effects for molybdenum determination (34).

Welz and Schlemmer (32,33) employed methane and Freon as modifiers for refractory elements. Methane was shown to increase the integrated absorbance signal for molybdenum, titanium, and vanadium and to reduce interferences for the determination of platinum group elements. The addition of Freon was shown to reduce memory effects for molybdenum with a lower clean step temperature, which extended the lifetime of the tube.

A list of recommended modifiers for each element is given in Table C.1. Stock solutions of most modifiers may be purchased from chemical companies or instrument manufacturers (see *Analytical Chemistry*, labguide edition). Alternatively modifiers may be prepared from ultrapure chemicals as described in Table C.4. It is usually necessary to optimize the chemical modifier employed for a given analysis by measuring the integrated absorbance as a function of pyrolysis temperature, as shown in Figure 5.3. It may also be necessary to optimize the amount of modifier used by examination of the temporal profile and integrated absorbance. Generally 1 to 50 μg of modifier are introduced.

5.2 CHEMICAL INTERFERENCES

Chemical interferences involve a nonanalyte concomitant that reacts with the analyte and prevents quantitative conversion of the analyte into gaseous atoms. Several types of chemical interferences have been described that depend on the process involved. *Volatile compound formation* occurs when an interferent reacts with the analyte to produce a more volatile analyte-containing molecule that is vaporized out of the furnace during the pyrolysis step. Chloride ions have been shown to cause chemical interferences for the determination of a number of elements, such as lead and thallium. *Involatile compound formation* involves the formation of analyte-containing substances that are incompletely vaporized at the maximum possible atomization temperature. Several elements, including tungsten, tantalum, and barium, form involatile carbides. *Gas-phase interferences* involve a chemical reaction between vaporized analyte, either as a compound or the metal, with a gas-phase concomitant to form a compound that does not dissociate in ground-state atoms. Chemical reactions that prevent atom formation that occur in a liquid (solvent) or solid phase are called *condensed-phase interferences*. It is usually difficult to classify the type of chemical interference present without specialized techniques.

Chemical interferences generally cause a suppression of the analytical signal because fewer gaseous atoms are formed. A decrease in the sensitivity (slope) is observed, as indicated in Figure 5.5. Perhaps the most troublesome and consequently most studied chemical interference involves chloride and volatile elements such as thallium or lead. Figure 5.6 shows a signal suppression induced by sodium chloride upon thallium (35).

Suggested mechanisms of this interference include a gas-phase interference (36,37), condensed-phase interference (35), more rapid expulsion of analyte due to the decomposition of chloride-containing concomitants (38), and trapping of analyte in chloride-containing crystals that do not decompose in the furnace (38). Mahmood and Jackson (39) recently demonstrated that for thallium, chloride interferences occur by a condensed-phase reaction at 500 to 700°C, probably producing thallium chloride.

Qiao and co-workers (35) investigated the action of palladium and magnesium nitrates to reduce chloride interferences for the determination of thallium. The analyte was reported to adsorb onto palladium, while the chloride

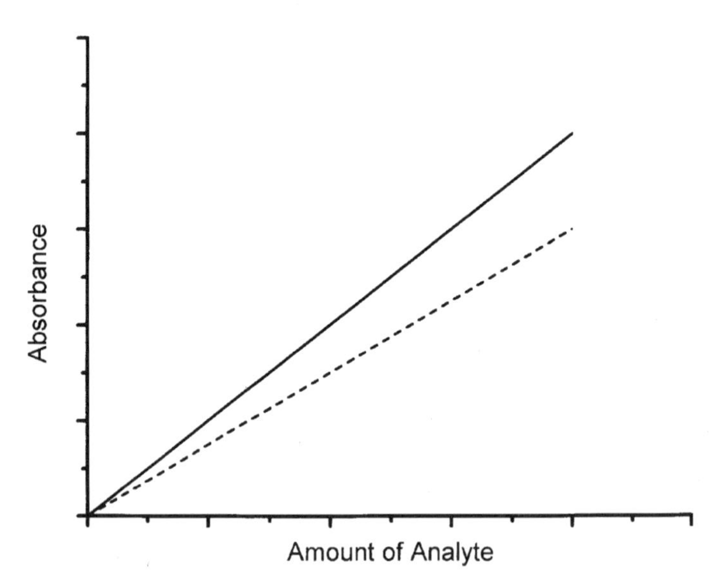

Figure 5.5. Effect of multiplicative (e.g., chemical) interferences upon a calibration graph: — no multiplicative interference and - - - multiplicative interference.

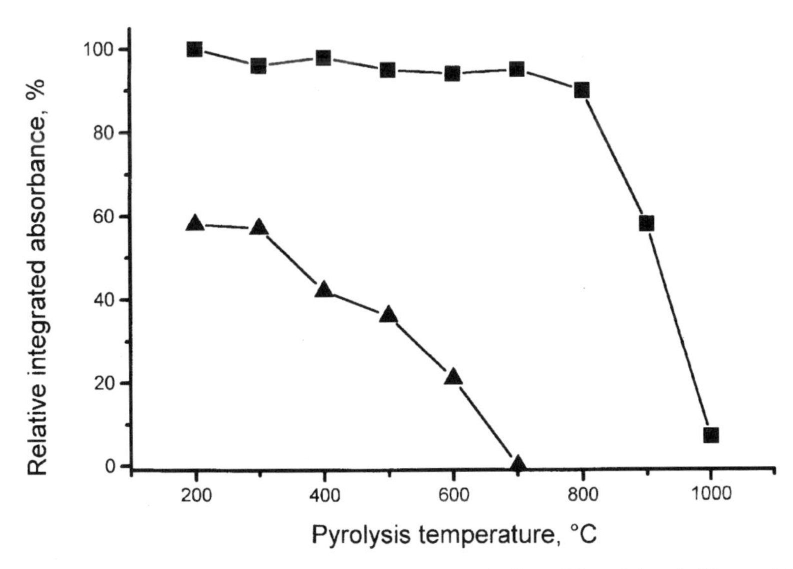

Figure 5.6. Thallium pyrolysis optimization graphs: 1 ng thallium (■) and 1 ng thallium and 10 μg sodium chloride (▲). Atomization temperature: 1900°C; wavelength: 276.8 nm (35).

was vaporized during the pyrolysis step. Hence interactions between thallium and chloride were avoided to eliminate this interference. After the chloride was removed, thallium diffused out of the modifier during the atomization step.

Regardless of the mechanism of the chemical interference, similar approaches can be used to minimize their effects. Several features of modern graphite furnace technology have been shown to be effective at reducing chemical interferences and allowing calibration with aqueous standards (Section 3.2). In general, the same type of approaches to minimize spectral interferences can be employed. First, the combination of a furnace with a high heating rate (>1000°C/s) and platform or probe atomization has been shown to reduce chemical interferences. This furnace design ensures the temperature of the graphite tube and the enclosed atmosphere have reached a high, constant temperature before atomization occurs, which reduces gas-phase reactions. Transversely heated furnaces heat more isothermally compared to longitudinally heated units and hence would be expected to have few chemical interferences (Section 4.2.1).

Second, chemical modifiers have been shown to reduce chemical interferences. For volatile elements, the role of the chemical modifier is to reduce the volatility of the analyte so that concomitants may be vaporized in the pyrolysis step prior to atomization. For many samples, a pyrolysis temperature of 1000°C is generally regarded as sufficiently high to vaporize enough matrix species to obtain accurate results, although the use of higher temperatures is desirable if the analyte is sufficiently stabilized. The use of palladium, either alone or with other metals (most commonly magnesium nitrate), has been shown to allow the use of pyrolysis temperatures ≥1000°C for most volatile elements (Table 5.3). Cadmium is a notable exception, with maximum temperatures of 700°C, but this temperature is the same or higher than temperatures obtained with other modifiers routinely used for this element.

Third, the use of pyrolytically coated graphite furnaces reduces chemical reactions between the analyte and the graphite. Although pyrolytically coated graphite may improve the analytical performance for volatile elements, it is essential for successful determination of carbide-forming elements. In the authors' experience, we have found that optimum performance for all elements is obtained with pyrolytically coated graphite.

In summary, modern furnace technology has been shown to minimize most chemical interferences. In some difficult matrices, the method of standard additions may be required to obtain accurate results.

5.3 PHYSICAL INTERFERENCES

Physical interferences involve changes in absorbance due to differences in the viscosity and surface tension of sample solutions. In general, changes in these solution properties have relatively little influence upon the quantity of material introduced in graphite furnace, but they can affect the degree of spreading of the sample on the graphite surface. Increased spreading of the sample may allow more interactions between the surface and the analyte, and consequently affect the efficiency of atomization.

One example of a potential physical interference involves slurry sampling, which involves the formation of an aqueous slurry of a powdered sample that is directly introduced into the tube (40,41). Surfactants are generally added to biological samples to assist in wetting and dispersing powders. In the absence of a surfactant, the material may clump and float on the surface. Surfactants may cause increased spreading on the graphite surface, which may increase interactions between the analyte and graphite. The use of pyrolytically coated graphite minimizes physical interferences because it is more impervious to solutions.

As a second example, other workers have employed bulking or thickening agents (e.g., glycerol) to prevent premature settling of particles in slurries. These chemicals have been reported to degrade the precision of pipetting by the autosampler (42).

REFERENCES

1. B. Welz, *Atomic Absorption Spectrometry*, 2nd ed., VCH, Weinheim, Germany, 1985.
2. W. Slavin and G. R. Carnrick, *CRC Crit. Rev. Anal. Chem.*, **19**, 95 (1988).
3. B. R. Culver and T. Surles, *Anal. Chem.*, **47**, 920 (1975).
4. S. Yasuda and H. Kakiyama, *Anal. Chim. Acta*, **84**, 291 (1976).
5. S. Yasuda and H. Kakiyama, *Anal. Chim. Acta*, **89**, 369 (1977).
6. K. Dittrich, *CRC Crit. Rev. Anal. Chem.*, **16**, 223 (1986).
7. F. Vajda, *Anal. Chim. Acta*, **128**, 31 (1981).
8. D. C. Manning, *Atomic Absorption Newsletter*, **17**, 107 (1978).
9. Y. Y. Zong, P. J. Parsons, and W. Slavin, *Spectrochim. Acta, Part B*, **49B**, 1667 (1994).
10. U. Heitmann, M. Schütz, H. Becker-Roß, and S. Florek, *Spectrochim. Acta, Part B*, **51**, 1095 (1996).
11. H. Massmann and S. Gucer, *Spectrochim. Acta, Part B*, **29B**, 283 (1974).
12. H. Massmann, Z. ElGohary, and S. Gucer, *Spectrochim. Acta, Part B*, **31B**, 399 (1976).
13. U. Kurfürst and J. Pauwels, *J. Anal. Atom. Spectrom.*, **9**, 531 (1994).
14. J. Y. Cabon and A. Le Bihan, *J. Anal. Atom. Spectrom.*, **9**, 477 (1994).
15. M. Sperling, B. Welz, J. Hertzberg, C. Rieck, and G. Marowsky, *Spectrochim. Acta, Part B*, **51B**, 897 (1996).
16. D. L. Tsalev, *Spectrochim. Acta Rev.*, **13**, 225 (1990).
17. W. Slavin, *Graphite Furnace AAS: A Source Book*, Perkin-Elmer Corporation, Norwalk, CT, 1984.
18. L. H. J. Lajunen, *Spectrochemical Analysis by Atomic Absorption and Emission*, Royal Society of Chemistry, Cambridge, England, 1992.
19. S. Xiao-quan and W. Bei, *J. Anal. Atom. Spectrom.*, **10**, 791 (1995).
20. W. Frech and B. V. L'vov, *Spectrochim. Acta, Part B*, **48B**, 1371 (1993).
21. G. Schlemmer and B. Welz, *Spectrochim. Acta, Part B*, **41B**, 1157 (1986).
22. J. M. Harnly and B. Radziuk, *J. Anal. Atom. Spectrom.*, **10**, 197 (1995).
23. P. Bermejo-Barrera, M. C. Barciela-Alonso, J. Mordela-Pineiro, C. González-Sixto, and A. Bermejo-Barrera, *Spectrochim. Acta, Part B*, **51B**, 1235 (1996).
24. J. Y. Cabon and A. L. Bihan, *Spectrochim. Acta, Part B*, **51B**, 1245 (1996).
25. I. L. Shuttler and H. T. Delves, *Analyst*, **111**, 651 (1986).
26. R. L. Bertholf and B. W. Renoe, *Anal. Chim. Acta*, **139**, 287 (1982).
27. H. T. Delves and J. Woodward, *Atom. Spectrosc.*, **2**, 65 (1981).
28. M. Beaty, W. Barnett, and Z. Grobenski, *Atom. Spectrosc.*, **1**, 72 (1980).
29. R. H. Eckerlin, D. W. Hoult, and G. R. Carnrick, *Atom. Spectrosc.*, **8**, 64 (1987).
30. G. Schlemmer and B. Welz, *Fresenius J. Anal. Chem.*, **328**, 405 (1987).

31. E. A. Nater, R. G. Burau, and M. Akeson, *Anal. Chim. Acta*, **225**, 233 (1989).
32. B. Welz and G. Schlemmer, *Atom. Spectrosc.*, **9**, 76 (1988).
33. B. Welz and G. Schlemmer, *Atom. Spectrosc.*, **9**, 81 (1988).
34. S. P. Ericson, M. L. McHalsky, and B. Jaselskis, *Atom. Spectrosc.*, **8**, 101 (1987).
35. H. Qiao, T. M. Mahmood, and K. W. Jackson, *Spectrochim. Acta, Part B*, **48B**, 1495 (1993).
36. G. F. R. Gilchrist, C. L. Chakrabarti, J. Cheng, and D. M. Hughes, *J. Anal. Atom. Spectrom.*, **8**, 623 (1993).
37. J. P. Byrne, M. M. Lamoureux, C. C. Chakrabarti, T. Ly, and D. C. Grégoire, *J. Anal. Atom. Spectrom.*, **8**, 599 (1993).
38. S. Akman and G. Döner, *Spectrochim. Acta, Part B*, **49B**, 665 (1994).
39. T. M. Mahmood and K. W. Jackson, *Spectrochim. Acta, Part B*, **51B**, 1155 (1996).
40. Z. A. de Benzo, M. Velosa, C. Ceccarelli, M. de la Guardia, and A. Salvador, *Fresenius J. Anal. Chem.*, **339**, 235 (1991).
41. N. J. Miller-Ihli, *Anal. Chem.*, **64**, 964A (1992).
42. N. J. Miller-Ihli, *J. Anal. Atom. Spectrom.*, **3**, 73 (1988).
43. H. Qiao and K. W. Jackson, *Spectrochim. Acta, Part B*, **46B**, 1841 (1991).
44. M. W. Hinds and K. W. Jackson, *J. Anal. Atom. Spectrom.*, **5**, 199 (1990).

CHAPTER 6

SAMPLE PREPARATION AND INTRODUCTION

Sample preparation involves the conversion of a sample into a form that is suitable for analysis, which in general is into a solution, although methods have been developed that allow the use of solids. In general, the quality and rate of GFAAS analysis is dependent on the success of sample preparation procedures (1,2). *Sample introduction* involves the transfer of a standard or prepared sample into the graphite furnace, and the method employed depends on the state of the sample after preparation. Due to problems with this process, Browner (3) has referred to sample introduction as the "Achilles' Heel of Atomic Spectroscopy." These two processes are closely related and hence will be considered together in this chapter.

Liquid, gaseous, and solid samples are all determined by graphite furnace atomic absorption, and here a general overview of sample preparation/introduction will be provided, along with representative applications. Conventional dissolution methods (acid digestion, combustion, fusion) as well as methods to analyze solids with minimal sample preparation (slurry and solid sampling) are discussed. Methods of preconcentration/isolation of analyte, such as extraction, chromatography, and flow injection, allow the removal of the analyte from its matrix and a reduction in the detection limit. A variety of applications of these methods have been employed with GFAAS. GFAAS has also been used to obtain quantitative information on the chemical form of metal present in samples, which is called metal speciation. A guide to the literature of GFAAS applications in Appendix B.5 of this volume provides literature reviewing the immense number of sample preparation procedures. A volume edited by Haswell (4) includes specific preparation/introduction procedures for a variety of samples by atomic absorption, including environmental, air, food, metallurgical, geochemical, ceramic, industrial, and forensic materials.

6.1 LIQUIDS

Aqueous samples (e.g., river water, seawater, etc.) can be introduced directly into the graphite furnace with the autosampler (Section 4.2.2). If the sample is viscous, such as blood, or colloidal (milk), then it is necessary to dilute the sample with an appropriate solvent. Usually deionized water or dilute nitric acid are employed for this purpose. Surfactants, such as Triton X-100, are added to some samples to lower surface tension and promote thorough mixing of the diluted sample. The use of a digestion procedure has been shown by some analysts to improve the detection limit and remove some interferences (5,6).

Some recent applications of waters and fluids are listed in Table 6.1. Van der Jagt and Stuyfzand (7) reviewed the merits and limitations of various methods, including GFAAS, for the determination of elements in water. The determination of lead in blood has been widely investigated due to the toxicity of the element, the relatively low concentration levels (typically 1 ng/mL in "normal" blood), and severe matrix effects. Deval and Sneddon (8) described a method for the direct, simultaneous determination of cadmium and lead in blood reference

Table 6.1 GFAAS Applications of Liquid Samples with Direct Sample Introduction or Simple Dilution

Sample(s)	Analyte(s)	Conditions	References
Blood	Cd, Pb	Direct introduction or dilution with deionized water	8
Seawater	Cu	Direct introduction	128
Seawater	Cd	Direct introduction	129
Water, urine, serum	Cr	Direct analysis (water, urine) or diluted with 0.2% Triton X-100 in 0.1% nitric acid (serum)	130
Urine, serum	Si	Dilution with deionized water	131
Reactor coolant water	Fe, Ni, Mn, Cr, Co	Direct introduction	9
Wine	As, Sb	Direct introduction; better detection limits with digestion; standard addition	6
Urine	Se	Dilution prior to digestion; matrix matched calibration	5
Serum	Ni	Dilution with 1% HNO_3 + 0.25% Triton X-100	132
Seawater	Zn	Direct introduction	133
Serum	Dy	Samples were diluted with 0.1% HNO_3 + 0.1% Triton X-100 and spiked with organic Dy	134
Serum	Ge	Dilution with 1*M* HNO_3; La modifier	135
Electrolyte solutions	Si	Dilution with deionized water	136

samples with self-reversal background correction. The use of an ammonium dihydrogen phosphate chemical modifier allowed the use of an elevated pyrolysis temperature that removed the blood matrix. A detection limit of 1.06 mg/mL was reported for lead, which allowed its direct determination at concentration levels between 4.8 and 17 ng/mL. Good accuracy was obtained by this method.

Tompuri and Tummavuori (9) analyzed reactant coolant water for five transition metals. Boric acid caused a background signal that interfered with these determinations. Hydrofluoric acid was added as a chemical modifier to remove boric acid from the sample during the pyrolysis step.

The concentration levels of metals in petroleum products (oils, gasoline, solvents, etc.) have been frequently determined by GFAAS (10). The usual method for calibration involves the use of organometallic standards available as a solution. These standards are available from the National Institute of Standards and Technology (NIST) (Gaithersburg, MD) (11) or from commercial vendors (e.g., Spex, Edison, NJ, or VHG Labs, Manchester, NH). The NIST standards, shown in Table 6.2, consist of 5 g of material that may be diluted to prepare working standards. Solvents employed for dilution include xylene, methyl isobutyl ketone, *n*-heptane, or kerosene. "Wear-metals in Oil" standard

Table 6.2 National Institute of Standards and Technologies Metalloorganic Compounds in Liquid Form (11)

Element	Metalloorganic Compound	Weight Percent	SRM
Barium	Barium cyclohexanebutyrate	28.7	1051b
Vanadium	Bis(1-phenyl-1,3-butanediono)oxovandium(IV)	13.01	1052b
Cadmium	Cadmium cyclohexane butyrate	124.8	1053a
Tin	Dibutyltin bis(2-ethylhexanoate)	22.95	1057c
Lead	Lead cyclohexanebutyrate	37.5	1059c
Lithium	Lithium cyclohexanebutyrate	4.1	1060a
Nickel	Nickel cyclohexanebutyrate	13.89	1065b
Silicon	Octaphenylcyclotetrasiloxane	14.14	1066a
Sodium	Sodium cyclohexanebutyrate	12.0	1069b
Strontium	Strontium cyclohexanebutyrate	20.7	1070a
Phosphorous	Triphenyl phosphate	9.48	1071b
Zinc	Zinc cyclohexanebutyrate	16.66	1073b
Aluminum	Aluminum 2-ethylhexanoate	8.07	1075b
Silver	Silver 2-ethylhexanoate	42.60	1077a
Chromium	Tris(1-phenyl-1,3-butanediono)chromium(III)	9.6	1078b
Iron	Tris(1-phenyl-1,3-butanediono)iron(III)	10.45	1079b
Copper	Bis(1-phenyl-1,3-butanediono)copper(II)	16.37	1080a

reference materials, which are also available from NIST (SRMs 1083, 1084a, 1085a) (11) may be employed to evaluate the accuracy of the analysis. Marshall (10) has described procedures for the direct analysis of metals in fuel oils, lubricating oils, and gasoline.

Saba and co-workers (12) determined aluminum, copper, iron, magnesium, and other critical wear metals in lubricating oils by GFAAS. The samples were diluted 1+4 with kerosene and directly injected into a furnace for analysis. This method was compared to atomic emission and flame AAS procedures and gave better sensitivity and precision than the other methods.

Direct analysis of petroleum and petroleum products is a relatively easy procedure but may give degraded accuracy because the metals present in oil may be in the form of suspended or colloidal inorganic compounds (10). The atomization efficiency may vary for the different chemical forms of the metals, resulting in degraded accuracy. In addition, problems with sample homogeneity may occur. In order to obtain the best accuracy and precision, it may be desirable to dissolve the sample by a wet ashing or a combustion method (Section 6.3) and calibrate using aqueous standards.

6.2 GASES

The deleterious effects of atmospheric and industrial pollution caused by metallic compounds in the air continues to be a major concern due to illness or death resulting from exposure (13–15). This is particularly true in the workplace. Legislation in the United States is controlled by the Occupational Safety and Health Administration (OSHA). The Environmental Protection Agency (EPA) maintains some control over the introduction of new and potentially hazardous materials into the industrial workplace. The American Conference of Governmental Industrial Hygienists (ACGIH) publishes a list of threshold limit values (TLVs) based on experiments on animals, reports of medical cases, and industrial experience. Threshold limit values are guidelines used in the practice of industrial hygiene and the control of potential hazards that are periodically redefined and are typically in the low to submilligram per cubic meter ranges for most metals. Reviews (13,16) on metal determination in air by various atomic spectroscopic techniques are available.

Sampling systems for airborne metals consist of three devices: a means of collecting an air sample, a device to trap the metal in air, and a means of measuring the amount of air sampled. While several methods have been proposed for collecting the air sample with various degrees of success, including sedimentation, centrifugal collection in cyclones, thermal precipitation, and condensation, the most suitable are filtration, impaction, and electrostatic precipitation.

At the current time, the most widely used and accepted method for collection of metals in air is filtration. A filter system consists of a sampling head, the filter, and a pump. Sampling heads are available in a number of sizes and are most commonly made with various plastics. The physical size and porosity of a filter influences the air flow; the flow rate depends on the pressure differential. In addition, the temperature decreases as the pores of the filter are blocked during sampling. Typical filter sizes are 25 or 47 mm diameter. Several manufacturers provide a full line and range of filter heads and filters that comply with EPA, OSHA, and National Institutes of Occupation Health Administration requirements (see *Analytical Chemistry Labguide*). Filters may be classified as depth or membrane filters.

Depth filters consist of cellulose or glass fibers giving tortuous channels of various sizes to trap particles. They have the advantage of high capacity and are inexpensive but have no defined pore size and low strength. Cellulose filters are inexpensive, tough, resistant to dilute chemical solutions, and easily decomposed by wet or dry digestion procedures. Glass filters have high mechanical strength and are resistant to moderate chemical attack, but may contain high concentration levels of metals.

Membrane filters are composed of a rigid continuous polymer with fine holes of uniform and defined size so that particles are held on the surface. They have the advantage of good separation of particles of different sizes and a high pore density that allows quick filtration. However, they have a low loading capacity and clog more quickly. They are composed of a variety of materials including polyvinylidine fluoride and Teflon, which are chemically inert and stable toward solvents, cellulose esters (acetate and nitrate), which are hydrophilic and readily combustible with negligible ash, or polyvinyl chloride, which is hydrophobic and has low water uptake. Teflon has the additional advantage of thermal stability up to 250°C.

The pump should have a calibrated flow rate indicator or total volume meter. The flow rate through a filter may change as it is clogged. Sampling time varies depending on the analyte concentrations but is typically from 30 min to several days or even weeks.

After sampling, the filter is removed and digested for GFAAS analysis (Section 6.3). Usually nitric acid is sufficient to remove the analyte from the filter, although other acids may be required as well (15). A common procedure for the analysis of gas samples by GFAAS involves the collection of airborne particles on a 0.8-μm pore size membrane filter through which air is pumped at a rate of 1.5 to 2.5 L/min (15). The exposure of an individual can be monitored with a personal sample collection system that monitors air in the breathing zone of the worker. The collection time depends on the concentration of the analyte. After sample collection, the filter is removed, digested with nitric acid (Section 6.3), and introduced as a solution.

Wang and co-workers (17) described a closed vessel–microwave digestion procedure in combination with various spectrometric techniques, including GFAAS, flame AAS, inductively coupled plasma optical emission spectrometry (ICP-OES), and inductively coupled plasma mass spectrometry (ICP-MS). The digestion procedure, using an acid mixture of $HNO_3/HClO_4/HF$ (1+3+7), provided good accuracy for all elements in a NIST urban particulate SRM with the exception of chromium. The authors suggested that the use of GFAAS, flame AAS, ICP-OES, and ICP-MS together may provide the most accurate analyses.

An alternative approach involves the direct *impaction* of air onto the surface of a graphite tube (18–20). The tube is then inserted into the electrodes of the furnace and a conventional analysis is performed. Lee et al. (18) described an improved version of a single-stage impactor with collection in a commercial graphite tube. The system was constructed from nylon to minimize metal contamination. Jets of 0.5, 1.0, and 1.5 mm inside diameter were constructed for optimization of the system. A turret was present in the system with positions for four graphite tubes. This allowed four consecutive, identical experiments to be performed.

The performance of this system was investigated for the determination of cadmium, chromium, lead, and manganese in uncontaminated air and air exposed to cigarette smoke. The precision varied within a factor of 3, which was expected due to inhomogeneity. A significant increase in the concentration levels of these metals was observed in air containing cigarette smoke compared to uncontaminated air. The authors suggested that the accuracy of the analyses was difficult to assess and concluded that the system was best suited to semiquantitative analysis. This impaction system provides a fast, convenient, sensitive method to simultaneously monitor four elements in a near real-time manner. Disadvantages include degraded precision, difficulty of calibration, and unsuitability for automation.

Electrostatic precipitation is a relatively new method for the determination of metals in air (21–25). Air is passed through a graphite tube that contains an electrode connected to a high voltage power supply. The application of approximately 2000 V produces a corona discharge that captures particles from the airstream.

Torsi and co-workers (25,26) have recently described an electrostatic precipitation system with a laboratory-constructed graphite furnace designed to provide high heating rates (8000°C/s). In addition, no dosing hole is present in the graphite tube. The electrostatic precipitation method was compared to a standard filtration method. The authors concluded that electrostatic precipitation provided higher efficiency than the standard method because approximately 30% of the particles were not trapped by the 0.45-μm filters. The special design of their graphite furnace system allowed absolute analysis (Section 3.5) without the use of aerosol standards. This is a particular advantage for air sampling because of the difficulty of preparing accurate standards.

6.3 SOLIDS

The majority of samples for analysis by GFAAS are solids, which are generally converted to solutions, and introduced in that form into the graphite tube. Some solids may be dissolved by simple dissolution in water [e.g., heroin (27)] but most require a digestion procedure. Most procedures involve the dissolution of 0.1 to 1 g of solid per 100 mL solution. Hoenig and Borger (28) described various dissolution methods for plant materials with flame and GFAAS analysis. The primary types of dissolution procedures are wet decomposition (acid digestion), combustion (dry ashing), and alkali fusion. The direct analysis of solids is also possible, and two basic techniques have been employed: slurry sampling, in which a powdered material is suspended in a solution that is aspirated into the atom cell, and solid sampling, in which a solid is directly inserted into the graphite furnace.

Sample preparation methods are generally considered to be critical to quantitative analysis because significant errors may occur due to loss of analyte due to volatilization or precipitation. Ohls (2) proposed that instrument companies work closely with specialty companies that produced sample introduction equipment. These specialty companies would provide standardized methods of sample preparation and introduction and serve as the link between sampling and analysis. The use of standardized methods of sample preparation would facilitate meaningful comparison of detection limits, linear dynamic ranges, and other analytical figures of merit between various spectrometers.

6.3.1 Wet Decomposition

Wet decomposition, or acid digestion, involves the use of mineral acids and oxidizing agents (hydrogen peroxide) to affect dissolution of a sample. Acid digestion is employed for a variety of organic and inorganic solids. Wet digestion procedures may be used to dissolve the entire sample (total decomposition), dissolve a fraction of the entire sample (strong attack), or simulate the transfer of elements in the environment, such as the assimilation of elements from soil by plants (moderate attack) (1).

Acids commonly used in these procedures include nitric, sulfuric, perchloric, hydrochloric, and hydrofluoric (Table 6.3) (1,29). Hydrochloric acid is usually avoided for furnace analysis to avoid chloride interferences (Section 5.2). Nitric acid generally serves as the primary oxidizing acid, and sulfuric acid is a dehydrating agent and has a high boiling point (300°C), which increases the rate

Table 6.3 Strong Acids Used in Wet Decomposition Procedures (1,29,115)

Acid	Applications	Limitations
HNO_3	Primary oxidant; may be used alone	Requires closed system for complete dissolution of organic material
H_2SO_4	Used with HNO_3 to raise boiling point; has oxidative and dehydrating properties; used with H_2O_2 to produce permono sulfuric acid	Cannot use large volumes in Teflon; may cause GFAAS interferences
$HClO_4$	Powerful oxidant with HNO_3	EXPLOSION HAZARD
HCl	Dissolution of inorganic salts	Nonoxidizing; causes severe interferences with volatile elements
HF	Required to dissolve silicates	Toxic; requires Teflon vessels; causes severe burns
Aqua regia (3+1 HCl and HNO_3)	Much more powerful than either acid alone due to formation of chlorine and nitrosyl chloride	HCl may cause interferences with volatile elements

of decomposition of some samples. The combination of hydrogen peroxide with sulfuric acid produces permono sulfuric acid in situ. Perchloric acid, although a potential explosion hazard, is a very strong oxidizing agent, and hence is typically mixed with nitric acid to reduce its reactivity. Hydrofluoric acid is required for the dissolution of silicates.

Total decomposition of most samples requires hydrofluoric acid combined with other acids. Strong attacks, which are performed with strong acids without hydrofluoric, are easier to use but will not dissolve silicate residues. This selectivity may be an advantage if the goal is to evaluate levels of pollution. Moderate attacks typically involve treating samples with dilute acids or other salts to evaluate the bioavailability of elements. For example, McLaughlin et al. (30) evaluated available aluminum, calcium, and magnesium concentrations in soil by extraction with $0.012M$ H_2SO_4 and $0.05M$ HCl. In other applications, it is desirable to monitor the concentration of a metal that is exchanged by a cation of an added salt solution (e.g., ammonium acetate, potassium chloride), which is called the exchangeable concentration (31).

Wet decomposition can be performed with either open or closed systems. Open systems may include Teflon beakers or test tubes in a shallow aluminum block on a hot plate. Open acid digestion is suitable for relatively "easy" samples (e.g., food and agricultural samples) and is relatively inexpensive but is unsuitable for some samples, relatively time consuming (1–24 h or more), and may allow evaporative loss of volatile elements.

Closed digestion systems allow pressures above atmospheric to be developed in the vessel. Higher pressures allow the acids to boil at higher temperatures and facilitate complete oxidation of the sample. Figure 6.1 shows that nitric acid must be heated to more than 200°C (which can only be achieved in a high-pressure closed digestion system) to completely dissolve biological samples. The efficiency of digestion is commonly evaluated by the residual carbon content, which is a convenient, quantitative measure. In addition, closed vessels eliminate loss of volatile elements and increase the rate of digestion.

Examples of closed digestion systems include a *decomposition bomb*, high-pressure asher, or a microwave digestion vessel. The former consists of a Teflon container surrounded by a stainless steel body. After introduction of the sample and reagents, the entire bomb is heated in a muffle furnace at temperatures up to 200°C. Higher temperatures may be achieved with a high-pressure asher (HPA) system, which is composed of a quartz digestion vessel mounted in an autoclave. Unlike the decomposition bomb, this system allows simultaneous monitoring of the temperature and pressure of the sample during the decomposition procedure. Several sizes of vessels are available (2–70 mL); the smallest fits directly on a Perkin-Elmer GFAAS autosampler, allowing analysis from the digestion vial.

Microwave digestion involves the use of 2450-MHz electromagnetic radiation to dissolve samples in a Teflon or quartz container (1,4,29,32–34). Microwaves interact with polar molecules and induce alignment of the molecular dipole moment with the microwave electric field. The field changes constantly, causing rotation of the molecules and intermolecular collisions, producing heat. Consequently, the rate of microwave digestion is dependent on the coupling efficiency of microwaves with mineral acids. Nitric acid has the highest efficiency, with a value nearly as high as water, followed by hydrofluoric and sulfuric acids.

Microwave ovens specific for chemical digestions are recommended for safety considerations. Both open and closed systems have been used with microwave digestion. Most dissolutions are performed with Teflon vessels because it is inert with respect to metals, although the maximum temperature to which they may be heated is 200°C. This property prevents the use of large quantities of sulfuric acid (boiling point 300°C), which may deform the vessels. To obtain higher temperatures, quartz vessels are employed. Closed systems (Fig. 6.2) allow faster digestions (<30 min), the digestion of more difficult samples (e.g., polymers, geochemical), and a reduced risk of analyte volatilization, but are relatively expensive. For example, a commercial

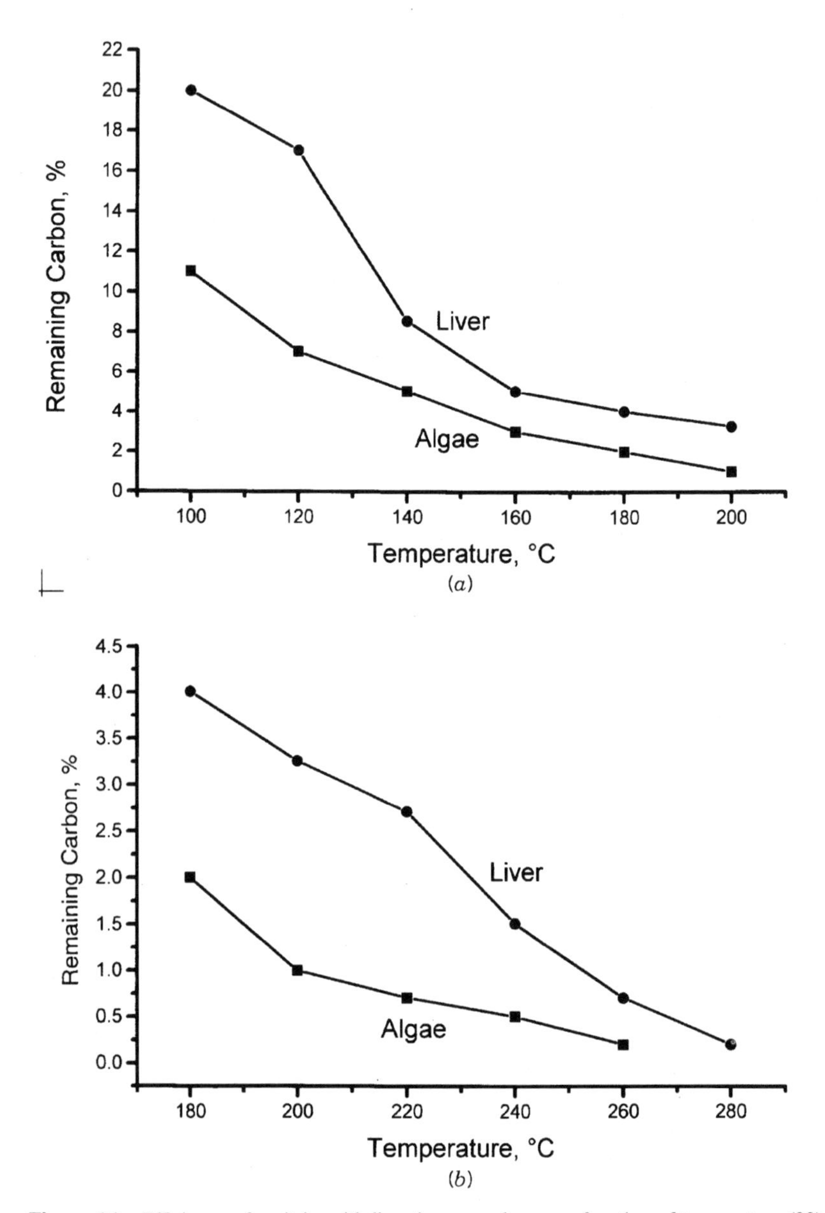

Figure 6.1. Efficiency of a nitric acid digestion procedure as a function of temperature (29).

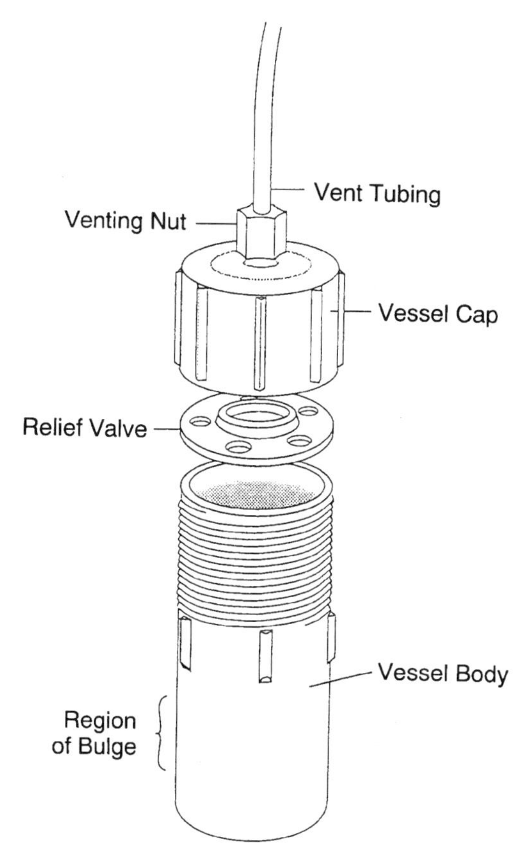

Figure 6.2. Commercial microwave digestion vessel (213). Reproduced with permission of CEM Corporation, Indian Trial, NC.

microwave oven is $15,000 to $20,000, and digestion vessels are approximately $100 each.

A particular advantage of microwave dissolution procedures is the ease of automation. A relatively new development is the combination of microwave digestion with on-line sample and reagent flow transport (35–38). These commercial instruments (CEM, Matthews, NC, and Questron, Mercerville, NJ) offer the potential for rapid, automated sample preparation. Solid samples are converted to 0.1 to 1% slurries by the addition of suitable acids. As with slurry sampling techniques discussed in Section 6.3.5, it is usually necessary to produce samples with a homogeneous, small particle size to ensure a uniform slurry. Some samples may require a "predigestion" step in order to prevent plugging of the instrument. Agitation of the slurries is performed by a paddle or ultrasound in order to produce a homogeneous slurry. An aliquot of slurry is obtained in a sampling loop and then pumped through the microwave system. An output autosampler is used to control the volume of digest delivered. Mineral acid solutions are placed in an acid-rinse reservoir to facilitate self-cleaning of the instrument.

Sturgeon and co-workers (37) evaluated the effectiveness of a commercial continuous-flow microwave system for the analysis of environmental National Research Council of Canada (NRCC) certified reference materials (CRMs) (lobster tissue, oyster tissue, and marine sediments). They reported that using "strong acid" attack on a closed vessel with microwave digestion at moderate pressures, typically 1 to 30% organic carbon remains present. The oxidation efficiency of the flow system for the digestion of lobster tissue was only 64%, but with one exception (chromium) the low efficiency did not degrade the accuracy of analyses performed by ICP-OES, ICP-MS, and GFAAS. The precision of the analyses was better than 1%.

Arruda et al. (35,36) determined aluminum and selenium in an NIST oyster tissue SRM and shellfish using a laboratory-constructed continuous-flow microwave system. The samples were ground to a particle size $<200\,\mu m$, suspended in 0.2% nitric acid, and homogenized in an ultrasonic bath before injection into the continuous-flow system. For aluminum, the recovery was approximately 90%. Accurate determination of selenium, which was present at low concentration levels, required a stoppage of the flow to ensure complete digestion of the sample.

In our opinion, the relatively high cost and moderate performance of these instruments makes their value questionable for GFAAS. In addition, it is necessary to homogenize and size-fractionate samples, as required with slurry sampling GFAAS. Slurry sampling accessories are more economical than continuous-flow digestion systems, and analysis can be performed directly after slurry formation.

Table 6.4 summarizes some recent digestion procedures employed in GFAAS. Virtually all procedures involve wet ashing, indicating the convenience of these methods compared to combustion (Section 6.3.2) and fusion (Section 6.3.3) methods. A number of microwave procedures have demonstrated the speed of these methods. However, several analyses were performed with traditional open-air digestion methods. Although these procedures require more time for digestion, it is reasonably easy to prepare a large number of samples (e.g., 100), digestion them, and prepare them for analysis. Consequently open digestion procedures will continue to be employed for routine analysis. Readers interested in more details on specific procedures for specific samples are referred to Appendix B.5 and references (4,29).

Table 6.4 Examples of GFAAS Digestion Procedures

Sample(s)	Analyte(s)	Digestion Procedure	References
Crayfish; sediments	Pb	Open-vessel hot-plate digestion with HNO_3 at 70°C; open-vessel digestion in HNO_3 at 70°C	137
Fish	Cd	Closed-vessel microwave digestion with $HNO_3 + H_2O_2$	138
Bismuth and tellurium oxides	Al, Cd, Co, Cr, Cu, Fe, Mn, Ni, Pb, V	Open-vessel hot-plate digestion with HNO_3 or HCl	139
Feedstuffs	Ca, Cu	Digestion in a Kjeldhal flask with $H_2SO_4 + H_2O_2$	140
Heroin; cocaine	Ag, Mn, Ni	Dissolution with water; digestion with HNO_3	27, 141
Bone	Pb	Closed-vessel microwave digestion with HNO_3	142
Milk	Al	Dry ash at 600°C; extract ash with 0.2%HNO_3; open-vessel hot-plate digestion with $HNO_3 + H_2SO_4$; dilution with 0.2% HNO_3; good accuracy reported with dilution procedure	143
Milk desserts	Al	Dry ash at 600°C; extract ash with 0.2% HNO_3 Open-vessel hot-plate digestion with $HNO_3 + H_2SO_4$ Automated system of slurry preparation	144
Cucumber plants	Pb	Closed-vessel microwave digestion with HNO_3	145
Soil; fungi	Cs, Ag	Dry ash at 450°C followed by closed-vessel microwave digestion with aqua regia; closed-vessel microwave digestion in $HNO_3 + H_2O_2$	146
Food	Pb	Open-vessel hot-plate digestion with HNO_3	147
Tin	Al	Open-vessel hot-plate digestion with HCl	148
Sediment; sewage sludge; leaves	Cr	Closed-vessel microwave digestion with aqua regia + HF + H_3BO_3; closed-vessel microwave	149

(*continued*)

Table 6.4 (*continued*)

Sample(s)	Analyte(s)	Digestion Procedure	References
		digestion with aqua regia + HF + H_3BO_3 + H_2O_2; closed-vessel microwave digestion with aqua regia + H_2O_2	149
Hair, nail	Se	Closed-vessel microwave digestion with HNO_3 + H_2O_2	150
Indium phosphide	Te	Open-vessel hot-plate digestion with aqua regia	151
Telluride thermoelectric material	Tl	Digestion with HNO_3	152
Nickel-based alloys; steel	Bi, Ag	Closed-vessel microwave digestion with HNO_3 + HF	153
Pharmaceuticals	Cd, Co, Cr, Cu, Fe, Mn, Ni, Pb, Sb, Sn, V	Dissolved in water	154
Sediments, flyash, biological samples	Cd	Closed-vessel microwave digestion with HNO_3 + HF + $HClO_4$	155
Bone and tissue	Si	Open-vessel digestion with HNO_3	156
Nickel-based alloys	Tl	Open-vessel digestion with HF + H_2SO_4 + H_2O_2	157
Sediment; lobster tissue, oyster tissue	As, Cr	Continuous-flow microwave-assisted digestion with HF + HNO_3 + HCl; continuous-flow microwave-assisted digestion with H_2O_2 + HNO_3	158
Shellfish	Al, Se	Continuous-flow microwave-assisted digestion with HNO_3	35, 36
Vegetable samples	Al, Ca, Cd, Cu, Fe, Mg, Pb, Zn	Evaluation of three wet digestion methods and a dry ashing method; a wet digestion procedure using HF and HNO_3 was deemed most suitable	158
Fly ash SRM	Tl	Pretreat sample with sulfuric acid; evaporate to dryness; fuse with $LiBO_2$	40

6.3.2 Combustion

Combustion (dry ashing) procedures involve heating a sample to a sufficiently high temperature (400–800°C) to remove the organic constituents. The traditional method involves placing the sample in a crucible (platinum or ceramic) or a test tube, followed by insertion in a muffle furnace for 1 to 6 h to induce quantitative decomposition and removal of organic material. The residue is composed of carbonates and oxides. The analyte is then extracted from the ash with a mineral acid (usually nitric acid for furnace work).

Dry ashing has the advantage of relative ease of use and allows decomposition of relatively large sample sizes followed by concentration in a relatively small volume of acid. However, it is a relatively slow process and is limited to relatively "easy" samples. Losses of volatile elements (Hg, As, Se, Cd, Pb, Tl) may occur. Losses of analyte may also occur due to retention of analyte in the ash. For example, nitric acid will not dissolve silica present in the ash.

Several examples of combustion methods are described by Vandecasteele and Block (129), including combustion in closed systems. In general, combustion methods have been replaced by wet decomposition procedures for most analyses by GFAAS, as demonstrated in Table 6.4.

6.3.3 Fusion

Fusion (39) procedures are well suited for the dissolution of samples that cannot be dissolved by other procedures (e.g., geological samples). The sample is mixed with a four to tenfold excess of a fusion reagent, which include alkali metal hydroxides, carbonates, or borates (e.g., lithium metaborate), and placed in a platinum or graphite crucible. The crucible is then inserted into a muffle furnace at 800 to 1000°C for 15 min to 6 h to form a molten salt. The melt is then poured in a dilute acid solution (usually nitric acid for GFAAS). The principal advantage of fusion is its applicability to nearly all samples. On the other hand, the large quantities of flux reagents may increase the blank level, making the technique unsuitable for many GFAAS analyses. In addition, volatile elements may be lost in the fusion step.

Grognard and Piolon (40) described a fusion method for the determination of thallium in a flyash standard reference by GFAAS. The addition and subsequent evaporation of sulfuric acid to the sample prior to the fusion procedure was shown to allow fusion with $LiBO_2$ at 950°C without loss of analyte. Good accuracy was reported by this technique.

6.3.4 Summary of Digestion Procedures

Here we attempt to provide the reader with some suggestions regarding digestion procedures for use with GFAAS (1,29), which are summarized in Table 6.5. First, wet digestion procedures are far more widely used for GFAAS analysis than combustion and fusion methods due to lower losses and easier automation. In particular, microwave digestion procedures offer the possibility of automated sample preparation. To evaluate total metal content, hydrofluoric acid should be used with other acids (e.g., nitric acid) to ensure dissolution of silicates. For most pollution studies, nitric acid with other acids (no hydrofluoric acid) is used to dissolve samples. In general, hydrochloric acid should be avoided when volatile elements are determined by GFAAS to minimize chemical interferences. More detailed procedures are available in the volume edited by Haswell (4) and the literature listed in Table 6.4 and Appendix B.

6.3.5 Solids Analysis with Slurry Sampling

An alternative to the dissolution of powdered samples is a technique called *slurry sampling* (41–47) in which the material is suspended in a liquid diluent. The liquid depends on the nature of the sample. For example, for most biological

Table 6.5 General Dissolution Procedures (1)

Sample	Digestion Goal	Dissolution Procedure
Soils, sediments, rocks	Total content	HF followed by HNO_3 or aqua regia
Soils, sediments, rocks	Strong attack (pollution studies) Silicates will not be decomposed	Aqua regia
Metals	Total content	Acid digestion
Plant material	Total content	HF with HNO_3
Plant material	Strong attack (pollution studies)	Wet digestion with $HNO_3 + H_2SO_4 + H_2O_2$
	Silicates will not be decomposed	
Animal tissues	Total content	Wet digestion with $HNO_3 + H_2SO_4 + H_2O_2$; volatile elements (Hg, Se, As should be dissolved in closed vessels to prevent volatilization losses)
Fly ash	Total content	HF followed by HNO_3 or aqua regia

and agricultural samples, the diluent is usually dilute (5%) nitric acid with a surfactant (e.g., Triton X-100) to ensure good wetting of the sample and to prevent the formation of clumps. The elimination of a dissolution step has the advantages of reducing analysis time and the probability of analyte loss during sample preparation. In addition, quantities of reagents are frequently decreased, which reduces the risk of contamination, and less sample dilution is required, which may lower the determinable mass of analyte. Compared to flame and plasma methods, GFAAS is well suited to slurry sampling because of the relatively long time that the sample remains in the atomizer (long residence times), which usually induces complete atomization of particles. Slurry sampling for GFAAS has recently been reviewed by Bendicho and de Loos-Vollebregt (44), de Benzo et al. (45), and Miller-Ihli (42).

In order to obtain precise and accurate results by slurry techniques, it is necessary to produce a homogeneous slurry. This usually requires the use of a mill to convert the sample into a powder with a small (<10 μm), homogeneous sample size. Typically 1 to 15 mg of powder are added per 5 mL of diluent. An effective method of agitating the slurry is also required. Methods of agitation include the stabilization of the slurry with a thickening agent (e.g., glycerol) or homogenization of the slurry. Stabilization has been shown to prevent accurate pipetting by the autosampler (48) and hence is not recommended.

Homogenization has been performed by use of a magnetic stir bar, a vortex mixer, the introduction of gas bubbles, high-pressure homogenization, or an ultrasonic probe. Magnetic stirring has the disadvantage that magnetic particles may adhere to the bar. Vortex mixing has been shown to be ineffective at producing homogeneous slurries of dense materials (e.g., sediments) and cannot be easily automated. The use of manual pipetting is inconvenient and also results in a degradation in precision (Section 4.2.2). Gas bubbling was shown to be ineffective at forming homogeneous slurries of samples in which the analyte was associated with larger particles.

Tan et al. (49,50) described a high-pressure homogenization procedure for slurry GFAAS of zoological and botanical samples. A sample (0.1 g) was diluted with 20 mL ethanol–water (1+9) containing 0.25% tetramethylammonium hydroxide (TMAH). The TMAH was added to help solubilize the tissues. The resulting mixture was macerated at 20,000 rpm for 60 s and processed through a 34.58-MPa homogenizer four times. Good accuracy was reported for the determination of several elements in SRMs, although the homogenizer induced lead and copper contamination. The slurries were reported to be stable up to 6 days after homogenization. This relatively new procedure is probably limited to relatively soft samples, and has not yet been automated.

Ultrasonic agitation appears to be the method of choice for slurry preparation. Ultrasound induces disaggregation, wetting, and dispersal of solid particles in a liquid. In addition, it has been shown to enhance extraction of analyte into the diluent. Miller-Ihli (46) and Carnrick et al. (47) described an automated ultrasonic mixing accessory for slurry sampling. This commercially available device (Perkin-Elmer, Norwalk, CT) consists of a titanium ultrasonic probe mounted above the autosampler that effectively agitates powdered samples. A gas-actuated cylinder is employed to control the vertical position of the probe, and its operation is synchronized with the autosampler. After the sample has moved directly below the autosampler arm, the probe moves into the sample cup, and the ultrasound is activated to produce a homogeneous sample. The probe is then lifted out of the sample and the ultrasound turned off. The autosampler arm then enters the cup, removes an aliquot, and dispenses it to the furnace. Schäffer and Krivan (51) described a modification of a commercial autosampler tray that allowed the preparation of 10-mL slurries (10–100 mg sample), instead of the 2-mL slurries (1–10 mg sample) prepared with standard autosampler cups. The use of larger slurry volumes improved the precision to 2 to 8% compared to 4 to 26%.

Table 6.6 lists some recent applications of slurry sampling. Miller-Ihli (52,53) described an international collaborative study on slurry sampling GFAAS. Samples were sent to 13 laboratories to determine lead and chromium in two NIST SRMs. Average recovery values were between 78 and 103%. The use of generally accepted furnace techniques was cited as critical to obtaining good accuracy. Inaccuracy was attributed to the use of miniflows during atomization, low chemical modifier masses, high pyrolysis temperatures, low atomization temperatures, small sample masses, and small autosampler volumes.

Slurry sampling has the potential for rapid analysis compared to dissolution procedures, but with some limitations. First, the measurement of small masses of sample (2–50 mg) is required, which is time consuming and may not be representative of the bulk sample.

Second, it is usually necessary to characterize the particle size and homogeneity of the sample, as well as the distribution of the analyte between the solid and liquid phases of the slurry. If a significant fraction of the analyte is extracted into the diluent, the analytical performance will be similar to a digestion. However, if the analyte remains associated with the solid, then the precision will probably be reduced compared to digestion procedures.

Third, careful optimization must be performed to obtain good results. Graphite furnace atomic absorption spectrometry parameters to be considered include pyrolysis and atomization temperatures, amount and type of chemical modifier, and oxygen ashing (Chapter 5). In addition, it is also necessary to characterize the sample in terms of homogeneity, density, and particle size. These parameters are used to optimize slurry preparation, as outlined by Miller-Ihli (54). It is generally assumed that at least 50 particles should be introduced in each 20-μL injection into the graphite tube. For a material with a density of 1 g/mL, 20 mg of sample is required per milliliter of diluent. Although accurate analyses have been performed with particle sizes exceeding 100 μm, it may be

Table 6.6 Slurry Preparations

Sample(s)	Analyte(s)	Slurry Procedure	References
Biological, agricultural, and coal SRMs	Al, Fe, Cu, Mn, Zn, Pb, Ca, Cr, Mg, Mo	Vortex and ultrasonic mixing; glycerol as a stabilizer; 5% nitric acid and 0.4% Triton X-100	48, 159, 160
Environmental and agricultural SRMs	Cr, Cu, Mn, Zn, Pb	Ultrasonic mixing with a commercial slurry sampler; 5% nitric acid and 0.005% Triton X-100	54
Sediments	Pb, Cr	International collaborative study; various techniques	52
Molybdenum oxide	Li		161
Food and agricultural SRMs	Tl, Mn, Pb	Vortex and ultrasonic mixing; 10–30 mg sample in 5% nitric acid and 0.04% Triton X-100	162
Glass materials	Co, Cr, Cu, Fe, Mn, Ni	Gas bubbling mixing; 1–50 mg sample in 1.75 mL 3% HF	163
Food and agricultural SRMs	Ag, Cu, Fe, Mn, Pb, Zn	Vortex mixing 5–15 mL 5% nitric acid and 0.04% Triton X-100	164
Soil	Ni, Cr	Samples were ground for 15 min; ultrasonic mixing; 0.075–0.15% m/v in 0.1*M* KOH, 0.05% nitric acid, 5% nitric acid	165
Lamp phosphor	Hg	Ground to average particle grain <15 μm; ultrasonic mixing; 10–40 mg of sample in 1 mL of 0.05% nitric acid	166
Coal and flyash	As, Pb, Tl	Commercial ultrasonic mixer; 1–15 mg sample in 1.5 mL 5% nitric acid and 0.04% Triton X-100	167
Soil and sediments	Cd, Pb, Se, Tl	Ground for 15 min; 0.005–10% w/v slurries in 5% HF (Cd, Pb); 40% HF (Se) or 30% HF (Tl)	168, 169

(continued)

Table 6.6 (*continued*)

Sample(s)	Analyte(s)	Slurry Procedure	References
Zoological and biological SRMs; liver and kidney	Cd, Cu, Pb, Cr, Cu, Fe, Mn, Ni, Se	High-pressure homogenization; 0.1 g in 20 mL ethanol–water (1+9) with 0.25% tetramethylammonium hydroxide	49, 50
Quartz and graphite powders	Al, Ca, Cr, Cu, Fe, K, Mg, Na, Ni, Pb, Si, Sn, Zn	Commercial ultrasonic mixer; 10–100 mg sample in 10 mL 0.1*M* nitric acid and 0.008% Triton X-100	51, 170
Titanium and zirconium dioxide	Si	Commercial ultrasonic mixer; 10–100 mg sample in 10 mL 0.1*M* nitric acid and 0.008% Triton X-100	171
Marine sediments	As, Cd, Hg, Pb, Sn	Ground to reduce particle size $<250\,\mu m$; agitated with zirconia beads in 3 mL water for 60 min to reduce particle size $<0.5\,\mu m$; diluted to 10 mL water and 0.1% Triton X-100; sample was stirred magnetically before measurement	172, 173
Leaves and hair	Cd, Cr, Mn, Pb	Ground to reduce particle size $<30\,\mu m$; ultrasonic mixing; 20–200 mg sample in 2 mL water	174
Milk desserts	Al	Homogenized in a blender manually for 2 min; 2.5 g sample in 25 mL 0.2% nitric acid with ultrasonic mixing; automated preparation module dispenses samples slurries into GFAAS autosampler cup	144
Boron	Cell suspensions	Cultured cells were detached by a trypsin–EDTA solution; centrifuged; separated from supernatant; treated with 9.5*M* HCl, 10% Triton X-100, and 5000 ppm Ca (modifier). The autosampler capillary was enlarged to prevent clogging.	175
Hair	Pb	Samples were washed, pulverized, sieved to a particle size $<125\,\mu m$; 0.1 g was added to 1 mL of water and stabilized with a thickening agent (Viscalex HV30 was optimum); stirred magnetically before sample introduction with a conventional autosampler	176

necessary to use a nonstandard autosampler capillary. The precision may be degraded as well.

In conclusion, in the authors' opinion, a dissolution procedure is probably preferable for routine analysis and relatively easy to dissolve samples. As discussed, a large number of samples can be analyzed with conventional digestion procedures. In addition, at the current time, only one manufacturer has an automated slurry sampling accessory. On the other hand, slurry sampling may significantly reduce analysis time for difficult-to-dissolve powdered samples (e.g., pharmaceuticals) and some environmental samples.

6.3.6 Direct Solid Sampling

Direct introduction of solid samples (*direct solid sampling*) (41–44) eliminates sample preparation procedures, which reduces analysis time and prevents contamination by reagents. In addition, there is no dilution of the sample, which allows measurement of lower levels of analyte than dissolution procedures. Typically a few milligrams of solid material are introduced into the furnace. Recent reviews of solid sampling were prepared by Langmyhr and Wibetoe (43), Bendicho and de Loos-Vollebregt (44), de Benzo and co-workers (45), and Miller-Ihli (42). Several systems specifically designed for solid sampling have been developed (44); three commercially available units are described here.

Carnrick et al. (55) described the use of a graphite sampling cup and tube for solids analysis called the cup-in-tube system, produced by Perkin-Elmer Corporation (Norwalk, CT) (Fig. 6.3). The sample is weighed directly into the cup (mass, 180 mg), which is then inserted into the tube. The cup is heavier than a conventional platform and blocks the sample from radiational heating. The primary heating mechanism is conduction through the wall of the cup, and hence vaporization occurs into a relatively cold atmosphere. The characteristic mass using the cup-in-tube is typically degraded compared to platform atomization.

A second system for solid sampling is the use of a graphite microboat (Fig. 6.4) (56), which is a flat piece of graphite (6×4 mm, mass 120 mg) with a cavity for the sample and a hole with which to transport it. The microboat, produced by Thermo Jarrell Ash (Franklin, MA), is inserted into a slot in a rectangular furnace designed for solid sampling. The microboat is in direct contact with the wall of the graphite tube, causing the microboat to heat up nearly as quickly as the wall. Hence atomization occurs into a relatively cool atmosphere, inducing vaporization interferences.

Kurfürst et al. (57) described an automated system for solid sampling that included a powder sampler, a microbalance, a transport and handling system, a GFAAS instrument, and a microcomputer for instrument control. This system is built in Germany by Grün-Optiks. A powdered sample is transferred onto a graphite boat by vibration. The boat is weighed and placed in a graphite tube for

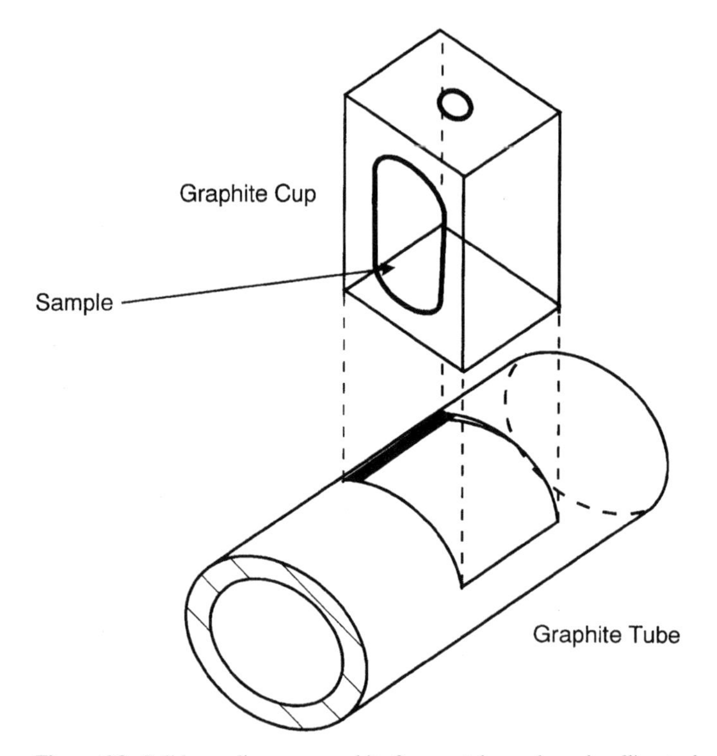

Figure 6.3. Solid sampling cup, graphite furnace tube, and cup-handling tool.

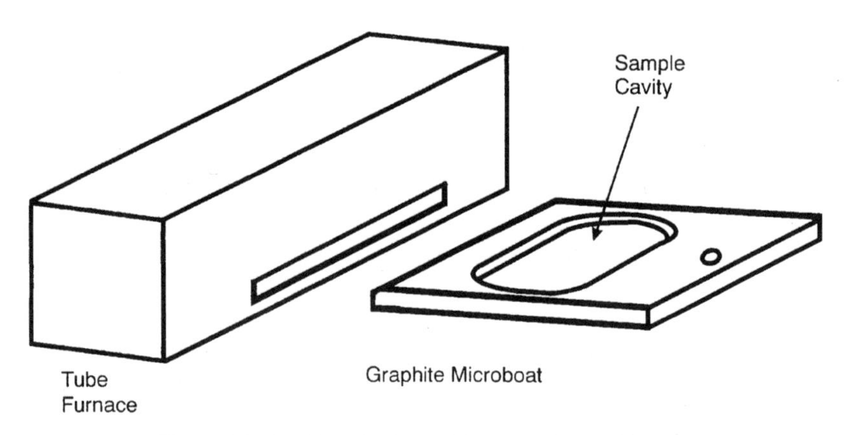

Figure 6.4. Graphite microboat with rectangular graphite tube.

analysis. This system would be expected to work well with homogeneous dry powders, but degraded precision and accuracy would probably result with wet, nonuniform samples.

Table 6.7 lists some recent applications of solid sampling. Van Dalen (58) described the use of a heated autosampler tray to reduce the viscosity of oils and fats and allow their direct injection into a graphite tube with an autosampler.

Table 6.7 Representative Applications of Direct Solid Sampling for GFAAS

Sample(s)	Analyte(s)	Solid Sampling Procedure	References
Blood, kidney, liver	Cd	Automated solid sampling system	59
Agricultural SRMs	Cd, Pb	Automated solid sampling system; comparison of uncertainty of digestion and solid sampling	60
Plastic film, PVC, liver	Cr; Pb; Cu	Cup-in-tube	55
Nickel-based alloys	Se	Microboat; better accuracy with digested samples	56
Nickel-based alloys	Bi, Pb, Se, Te, Tl	Cup-in-tube	62
Blood	Pb	Pipette or weigh blood on platform; comparison with digestion procedure and flame AAS	177
Silver	Au, Pd, Pt	Cup-in-tube; comparison to a slurry method	178
Gold	Si	Cup-in-tube; comparison to a digestion procedure by GFAAS and ICP-OES and spark ablation ICP-OES	179
Coal, flyash, and urban particulate matter	Cd, Ni	Cup-in-tube	180
Oils and fats	Cd	Autosampler tray heated samples to 60°C to reduce viscosity; conventional pipetting by autosampler	58
Tantalum	Cu, Fe, Mn, Na, Zn	Weigh samples on platform	181
Molybdenum	Cu, K, Mg, Mn, Na, Zn	Vaporize solid samples from cup furnace into a tube furnace or from a platform	182
Animal feeds	Cu	Weigh onto platform	183

Lücker and Schuierer (59) evaluated error sources in the determination of cadmium in tissue samples by the automated solid sampling system. They recommended the use of masses exceeding 0.07 mg in order to obtain optimum precision and in general concluded that mass measurement is a minor contributor to errors associated with solid sampling. The automated system was reported to reduce manual working time by 80%.

Kürfurst et al. (60) compared the uncertainty for the determination of cadmium and lead in agricultural SRMs by the automated solid sampling system with a digestion procedure. They concluded that different types of uncertainty contribute in the two methods of calibration but that the overall uncertainties were of comparable size. Berglund and Baxter (61) developed a computer package (SOLIDS) to perform statistical analysis of solid sampling GFAAS data.

In the authors' opinion, solid sampling is most applicable for materials that are obtained as 0.5- to 5-mg chunks or chips (e.g., metals, plastics). Dissolution or the slurry technique is preferable for most powdered samples. Solid sampling has a number of disadvantages. First, an automated system for the analysis of chips has not been developed. Second, the sample cannot be easily diluted, which is a problem because of the relatively short linear range of GFAAS. The conventional method involves mixing the sample with an inert powder, such as graphite. An alternative wavelength (if available) can be used to reduce the sensitivity. The use of miniflows is generally avoided because of degraded GFAAS performance. Third, the milligram masses used may be unrepresentative of the bulk properties of the sample. Fourth, it may be difficult to completely atomize the sample, and high backgrounds are often produced, and hence accurate analysis with aqueous standards may not be possible. It is often difficult to incorporate chemical modifiers, which are required in many analyses for accuracies, although for some metal samples (e.g., nickel-based alloys), the matrix may serve as the modifier (62). Fifth, the precision of solid sampling is usually degraded compared to dissolution procedures. Lastly, as in the case of slurry sampling, optimization of the instrument conditions is critical to obtain accurate and precise results. In general, we believe that the limitations of solid sampling outweigh its advantages for most analyses, at least until an automated system for the analysis of sample chips is available.

6.3.7 Laser Ablation

Laser ablation (LA) (63) involves the use of a laser beam to ablate, or vaporize, a solid sample. Spectroscopy may be done in the plasma generated by the laser beam or the vaporized sample may be transported to a conventional atom cell (63). The majority of work in this area involves the use of an inductively coupled plasma (ICP) as the atom cell, with detection by either optical emission spectrometry (OES) or mass spectrometry (MS) (64–67). The popularity of LA-ICP methods is due to the low transport efficiency (~1%) of conventional nebulization methods of sample introduction. Laser ablation allows a significant improvement in efficiency. Of course, the efficiency of GFAAS is 100%, and hence no gain in efficiency will be achieved by LA. Also ICP-OES and ICP-MS have the advantage of being multielemental techniques, compared to GFAAS, which until the early 1990s was almost exclusively single-elemental. However, laser ablation has the ability to vaporize microparticles (e.g., individual crystal grains in minerals), which cannot be achieved with other solid sampling techniques. For this reason, a couple of recent applications of LA-GFAAS have been reported.

Zhou et al. (68) described an LA-GFAAS setup that employed a 930-nm, 50-W continuous-wave diode laser. This laser was used to vaporize dry bovine liver powder. The best precision was obtained by the impaction of vaporized sample through a pipette tip into a graphite tube. The detection limit was estimated to be 0.1 ng, and the precision was 12%. It was suggested that an improvement in analytical performance would be obtained by a higher power laser system. In addition, the collection efficiency of lead by the impaction system was a relatively low 15%.

Wennrich and Dittrich (69) used a ruby laser for the determination of silver, cadmium, iron, manganese, nickel, lead, and zinc by LA-GFAAS. Two LA methods were compared: (1) a deposition method in which the sample was ablated, transported to a conventional graphite furnace system at room temperature, and then exposed to a conventional heating program and (2) an injection method in which the sample was vaporized into a constant-temperature furnace maintained between 1800 and 2500°C. The former system gave better sensitivity and precision for the more volatile elements (cadmium, lead, and zinc), but the latter was more sensitive for the other elements.

Although more development is required to develop standard accessories and methods, LA-GFAAS has considerable potential for microsampling. Commercial ablation cells are available for ICP techniques that can be used for GFAAS. One problem remains the relatively low collection efficiency of metals by impaction methods. The use of electrostatic precipitation (Section 6.2) may be useful in this regard.

6.4 PRECONCENTRATION/SEPARATION METHODS

The levels of elements in some samples (e.g., semiconductor industry, environmental samples) are below the detection limits attainable by graphite furnace atomic absorption. Some matrices (e.g., silicates) cause significant degradation of detection limits, and hence separation of the analyte from the matrix is required. Preconcentration/separation techniques are used to increase low levels of analyte and remove the sample matrix from the analyte. The enrichment factor (*E*) is a quantitative measure of the degree of preconcentration. It is defined as the concentration of analyte after the preconcentration step divided the analyte concentration in the original solution.

Commonly used methods for preconcentration/separation include extraction and chromatography. Major disadvantages of these preconcentration techniques include their labor-intensive nature and unsuitability to automation. Recently, flow injection (FI) has been employed with graphite furnace AAS, which has allowed rapid, automated preconcentration/separation procedures. Other methods include the use of biological organisms for preconcentration, coprecipitation, solid sorption, and a liquid supported membrane.

6.4.1 Extraction

Extraction (29,70,71) procedures involve the transfer of the analyte from a solvent (usually water) to a second solvent (usually organic). In order to obtain quantitative extraction, it is necessary to ensure that the analyte is all in the same chemical form, usually as its most common cation, and to control the pH of the aqueous phase. The limit of detection is generally improved by a factor of 10 to 20 by extraction methods, with a maximum enrichment of approximately 100. The analyte must be converted to an uncharged compound or to an ion association complex in order to increase its solubility in an organic solvent. The extraction process is evaluated in terms of the *distribution ratio* D (29)

$$D = \frac{C_{A,\mathrm{Org}}}{C_{A,W}} \tag{6.1}$$

where $C_{A,\mathrm{Org}}$ and $C_{A,W}$ are the concentrations of analyte in the organic and aqueous phase, respectively. The mass of analyte remaining in the aqueous phase after n extractions ($m_{A,W,n}$) is given by:

$$m_{A,W,n} = \left(\frac{V_W}{DV_{\mathrm{Org}} + V_W}\right)^n m_{A,W,0} \tag{6.2}$$

where $m_{A,W,0}$ is the initial mass of analyte in the aqueous phase and V_W and V_{Org} are the volumes of the aqueous and organic phase, respectively. In general, it can be assumed that quantitative transfer of analyte may be achieved in one step for D values exceeding 100.

Komárek and Sommer (72) reviewed organic complexing agents employed for atomic absorption, which are classified as metal chelating agents or ion association complexes. Metal chelating agents, such as 8-hydroxyquinoline (oxine) or ammonium pyrollidine dithiocarbamate (APDC), are one class of compounds used to remove analytes from a sample matrix. These chelating agents are commonly weak acids designated by HR. They can be used for a wide variety of metals (M^{n+}) (29,70,71), as illustrated in Table 6.8.

The equilibrium for the extraction process for a chelate may be expressed as:

$$M^{n+}{}_{(aq)} + n\mathrm{HR}_{(org)} \leftrightarrows \mathrm{MR}_{n(org)} + n\mathrm{H}^{+}{}_{(aq)} \tag{6.3}$$

and the distribution ratio for a metal–chelate system is given by (29):

$$D_M = \beta_n K_{ex} \frac{[\mathrm{HR}_{\mathrm{Org}}]^n}{[\mathrm{H}^+]^n} \alpha_M \tag{6.4}$$

where K_{ex} is the overall extraction constant; β_n is the formation constant of the metal chelate; and α_M is the fraction of the total metal concentration present as the uncomplexed metal. Equation (6.4) demonstrates that the distribution ratio decreases as the pH decreases. Variation of the pH can therefore be used to control the metal ions extracted.

The *$pH_{1/2}$ value*, which is defined as the pH at which 50% of a metal is extracted, is used to evaluate the selectivity of an extraction (Table 6.9). In general, a difference of three units in $\mathrm{pH}_{1/2}$ value is required to quantitatively separate two metal ions using a single batch extraction. For metals that cannot be separated on the basis of pH, additional complexing agents called masking

Table 6.8 Examples of Chelate Extraction Systems (29)

Chelate	Metals extracted
α-Dioximes (e.g., dimethylglyoxime)	Ni, Pd
1-(2-pyridylazo)-2-napthol (PAN)	Cd, In, Mn, Pd, U(VI), V(V), Zn, and others
1-Nitroso-2-napthol	Co(II)
8-Hydroxyquinoline (oxine)	Al, Mg, Sr, V, W, and others
Ammonium salt of *N*-nitrosophenylhydroxylamine (cupferron)	Fe(III), Ga, Hf, Mo(VI), Sb(III), Sn(IV), Ti(IV), U(IV), V(V), Zr, and others
Dialkyldithiocarbamates	As(III), Bi, Sb(III), Se(IV), Sn(IV), Te(IV), Tl(III), V(V) and others
Diphenyldithiocarbazone (dithiozone)	Ag, Bi, Cu, Hg, Pb, Pd, Pt, Zn, and others
β-Diketones (e.g., acetylacetone)	Alkali metals, Be, Sn, Cr, Mn, Mo, and others

Table 6.9 Values of $pH_{1/2}$ for Metal Dithiazonate Complexes (29)

Metal	$pH_{1/2}$
Ag	1.0
Bi	2.5
Cd	11.6
Cu	1.9
Hg	0.3
Pb	7.4
Sn	4.7
Tl	9.7
Zn	8.5

agents may be employed. *Masking agents*, which include EDTA and ammonia, serve to tie up one of the metals and prevent its extraction into the organic phase.

Ion association complexes involve the formation of a soluble ionic compound containing the analyte and a suitable counterion. In order to be suitable for extraction, these complexes should have no net charge or include sufficient nonpolar functional groups to allow high solubility in nonpolar solvents. Examples of ion association complexes include $Fe(o\text{-phen})_3(ClO_4)_2$ (where *o*-phen = ortho-phenanthroline) and $[(C_2H_5)_2O]_3H(H_2O)_n(FeCl_4)$.

Some examples of recent extraction procedures employed in conjunction with GFAAS are shown in Table 6.10. Extraction methods are relatively simple and may allow extraction of several elements or only one depending on the analytical requirements. However, these procedures are difficult to automate, relatively labor-intensive, and have interferences that reduce the extraction efficiency.

6.4.2 Chromatography

A variety of chromatographic procedures (29,73) have been employed for preconcentration/separation. Although typical enrichment factors of 100 are obtained, concentration factors up to 2000 have been reported. In order to obtain good accuracy, it is necessary to convert the analyte into one chemical form. A prescribed volume of sample is loaded onto a column using a mobile phase that does not elute the analyte but (ideally) allows removal of the matrix. A second mobile phase is then added that serves to quantitatively and rapidly remove the analyte from the column, resulting in a relatively concentrated solution. Ebdon, Hill, and Ward have reviewed the direct coupling of gas chromatography (GC)

Table 6.10 Extraction Procedures Employed with GFAAS

Sample(s)	Analyte(s)	Extraction Procedure	References
Urine	Sb	Extracted with cupferron in 12*M* HCl into methyl isobutyl ketone	184
Sediment porewater	Ag, Cd, Co, Cu, Fe, Ni, Pb, An	Extraction with ammonium 1-pyrrolidine dithiocarbamate and diethylammonium diethyldithiocarbamate into chloroform	185
Copper–nickel ores	Au, Ir, Os, Pd, Pt, Rh, Ru	Extraction with alkylaniline hydrochloride (0.4*M*) and petroleum sulfides (0.15*M*) into toluene	186
Water	Se	Extraction with sodium diethyldithiocarbamate into chloroform	187
—	As, Bi, Cd, Pb, Sb, Se, Sn, Te	Extraction with ammonium pyrrolidine dithiocarbamate into methyl isobutyl ketone; palladium or platinum were extracted as modifiers using methyltrioctylammonium chloride	188
—	As, Sb, Se, Te	Add $TiCl_3$ as a reducing agent; extraction with ammonium pyrrolidinedithiocarbamate into $CHCl_3$–CCl_4 (1+1)	189
—	Cd, Cu, Fe, Ni, Pb, Zn	Extraction with ammonium pyrrolidinedithiocarbamate and diethylammonium diethyldithiocarbamate into either $CHCl_3$, Freon, or methyl isobutylketone; back extract with nitric acid	190
Biological material	Gd	Extraction with 4-benzoyl-3-methyl-1-phenyl-2-pyrazolin-5-one in pH 6.7 ammonium acetate buffer into methyl isobutyl ketone	191

(74) and liquid chromatography (LC) (75) with atomic spectrometry, including GFAAS.

Ion-exchange techniques have been employed with cationic resins, which contain acidic functional groups (e.g., $—SO_3H$, $—CO_2H$) and anionic resins, which contain basic functional groups ($—NR_3H$). The exchange processes for cationic resin with a metal M^{n+} and anionic resin with an anion A^{n-} are:

$$n\mathrm{RSO_3}^-\mathrm{H}^+{}_{(s)} + \mathrm{M}^{n+}{}_{(\mathrm{aq})} \leftrightarrows (\mathrm{RSO_3}^-)_n\mathrm{M}^{n+}{}_{(s)} + n\mathrm{H}^+{}_{(\mathrm{aq})} \tag{6.5}$$

and

$$n\mathrm{RNR_3}^+\mathrm{X}^-{}_{(s)} + \mathrm{A}^{n-}{}_{(\mathrm{aq})} \leftrightarrows (\mathrm{RNR_3}^+)_n\mathrm{A}^{n-}{}_{(s)} + n\mathrm{X}^+{}_{(\mathrm{aq})} \tag{6.6}$$

The distribution coefficient for ion exchange (D_c) is given by

$$D_c = \frac{\text{concentration of analyte in the resin (mass/kg dry resin)}}{\text{concentration of analyte in solution (mass/L solution)}} \tag{6.7}$$

Cationic resins can be used to preconcentrate metal cations, while anionic resins allow the removal of negatively charged interferents and separation of the analyte if an anionic complex of the metal is formed (e.g., $ZnCl_4^{2-}$).

Chelating ion-exchange resins, such as Chelex-100, include a functional group that forms chelates with metal cations and have the advantage of forming stronger complexes with most transition-metal cations. Chelex-100 and other chelating ion-exchange resins are well suited to seawater samples because they do not interact significantly with alkali metal cations (Na^+), which may interfere with conventional ion-exchange resins. A more recent development is the use of preconcentration with a reversed-phase liquid chromatography procedure. A conventional reversed-phase column is used to separate metal ions that have been treated with a chelating agent.

Three GFAAS applications involving chromatographic preconcentration methods are listed in Table 6.11. We have included a preconcentration method that involves sorption onto carbon, which although not prepared as a chromatography column, could be adapted to that configuration. In spite of the high enrichment factors possible, this technique is relatively slow and is not well suited to automation. For most applications, chromatography has been replaced by flow injection procedures, which can be automated to be synchronized with the operation of a GFAAS instrument.

Table 6.11 Chromatography Preconcentration Procedures Coupled with GFAAS

Sample(s)	Analyte(s)	Chromatography Procedure	References
Seawater	Cd, Co, Cu, Fe, Mn, Ni, Zn	Formation of metal-8-hydroxyquinoline complex; metals separated on C_{18}-bonded silica gel	192
Titanium oxide	Ca, Cd, Co, Cr, Cu, Fe, K, Li, Mg, Mn, Na, Ni, Pb, Tl, V, Zn	Sample was digested with HF; analytes passed through a column of the cation exchanger Dowex 50WX in 0.1*M* HF; elution was performed in opposite flow direction with 4*M* nitric acid	193
Water, soil	Ag	Analyte complexed by ammonium salt of dithiophosphoric acid *O*,*O*-diethyl ester and sorbed on to carbon; analyte was desorbed by nitric acid	194

6.4.3 Flow Injection Analysis

Flow injection (FI) (76–82) analysis involves the introduction of a sample (typically 50 μL) in a flowing stream of liquid (~1 mL/min) in narrow-bore (0.5 mm), nonwetting tubing for quantitative analysis. A peristaltic pump is generally used to transport the liquid in a laminar flow pattern. A detection system, which is used to measure the analyte concentration, may be virtually any instrument. An autosampler is often used to inject the samples into the flow stream. A variety of types of chemical processes may be automated by FI. For example, a column, extraction module, or dialysis module may be used to separate the analyte from other sample constituents to minimize interferences in the detection system. Alternatively, reagents may be injected into the system to react with the analyte. It is desirable to ensure good mixing by coiling the tubing tightly or packing the tubing with beads to produce a packed-bed reactor.

The degree of mixing of the sample with the flow stream is referred to as *dispersion* (D). The dispersion of an FI system is usually quantified by the ratio of the analyte concentration injected (C_0) to the analyte concentration at the peak maximum (C_P):

$$D = \frac{C_0}{C_P} \tag{6.8}$$

The dispersion of an FI system may be controlled experimentally by variation of several design parameters. For example, dispersion increases with tubing length, tubing diameter, and the flow rate, and decreases with volume injected, tight coiling, and packing with glass beads.

As discussed above, the ability of a preconcentration system to increase the analyte concentration may be expressed by the enrichment factor *E*. However, in some cases an FI system is operated under different experimental conditions than a batch system, which may lead to an increase in sensitivity. Consequently, if experimental conditions are not identical between a batch and an FI system, the increase in sensitivity should be referred to as an *enhancement factor*.

Compared to conventional batch procedures in which each sample is located in a separate vessel (e.g., extraction), FI is a continuous-flow technique in which a series of samples are injected into a length of tubing separated by solvent. The basic processes of loading and removal from a column are similar to those employed in chromatography, but FI is distinguished from chromatography because it is designed for rapid quantitative analysis of a limited number of analytes instead of the separation of any number of compounds.

Flow injection has been widely employed with flame AAS as a method of preconcentration since the early 1980s because of its compatibility with a continuous-flow system. The combination of FI with GFAAS did not occur until the late 1980s, but since then a number of applications have appeared in recent years. The interest in FI-GFAAS may be related to the ability to do automated preconcentration steps and to the availability of a commercially available FI system (Perkin-Elmer, Norwalk, CT) for use with atomic absorption instruments. In general, the combination of GFAAS with FI for preconcentration requires specific features in terms of the instrument design (80,81).

First, GFAAS operates in a batch mode, and consequently preconcentration of the analyte is performed in parallel and must be synchronized with the atomization cycle in a discontinuous manner. Second, the maximum volume that can be accommodated in a graphite tube is less than 100 μL; this value is reduced with a platform and organic solvents that are commonly used for elution. When preconcentration is achieved by chromatography, it is therefore necessary to use microcolumns (15 μL), and it may not be possible to collect all of the analyte. Third, GFAAS is relatively sensitive to high concentrations of matrix elements, and hence it is usually necessary to incorporate a column washing step to remove residual sample matrix before elution of the sample. Fourth, the combination of these specifications results in a relatively complicated elution sequence that generally must be computer controlled.

An example of an FI manifold and its automated operation for GFAAS for the preconcentration of lead is shown in Figure 6.5 and Table 6.12. Lead was converted on-line to a dithethyldithiocarbamate chelate that was preconcentrated on a 15-μL microcolumn (Fig. 6.5, Table 6.12, step *c*), constructed from an

Eppendorf pipette tip and packed with C_{18} sorbent. In this study, deionized water was used to wash the matrix species from the column before introduction into the furnace (step *d*). More recent studies demonstrated that dilute nitric acid removes a higher fraction of the matrix compared to deionized water (83). The lead chelate (75 μL) was collected in the furnace autosampler capillary, which served as a collector tube (step *e*). A relatively low flow rate (0.15 mL/min) was used to allow equilibrium to be approached during this step, also to minimize dilution. It was not possible to collect the entire elution volume (~200 μL) because of the relatively large volume, and hence the most concentrated fraction was collected. A stream of air was then used to introduce the eluate into the furnace (step *a*). Residual analyte was eluted from the column to waste by ethanol (step *b*). The flow injection procedure increased the concentration of lead in solution by a factor of 26.

Several methods of separation have been employed with GFAAS, including sorption (84), ion exchange (85), extraction (86,87), coprecipitation (88,89), supported liquid membrane (90), and electrochemistry (91). The combination of flow injection methods for digestion is discussed in Section 6.3.1. Table 6.13 lists some applications of FI-GFAAS.

Sorption methods have been the most commonly used preconcentration method for GFAAS. Sperling et al. (84) described an automated FI procedure for the determination of lead using a 50-μL microcolumn packed with a highly selective macrocycle immobilized on silica gel (Pb-02). After the sample was loaded and washed with 0.15*M* nitric acid, residual solution was removed from the system with a flow of air. The analyte was eluted quantitatively into a graphite tube with 0.035*M* EDTA solution at a pH of 10.5. An enrichment factor of 77 was reported with a sample throughput of 17/h.

Bäckström and Danielsson (86,87) described an on-line solvent extraction system. Several elements were extracted as dithiocarbamates into Freon 113 and reintroduced into aqueous solution by a dilute mercury(II) solution. Enrichments of 50 to 100 were reported with a sampling frequency of 30/h.

Malcus and co-workers (90) described an automated FI system that employed a supported liquid membrane for preconcentration of aluminum, cadmium, and copper. Four porous, hydrophobic membranes were operated simultaneously in parallel. The sample was pumped into the donor side of the membrane, where the analytes diffused through the membrane to induce enrichment on the acceptor side of the membrane. The acceptor solution was then eluted into the GFAAS autosampler. The membrane system then initiated a second enrichment procedure during the GFAAS analysis of the first set of samples. An enrichment time of 30 min gave enrichment factors of approximately 10.

An on-line electrochemical preconcentration system was described for the determination of manganese (91). The analyte was deposited on either the cathode or anode, depending on the applied voltage, and elution was performed

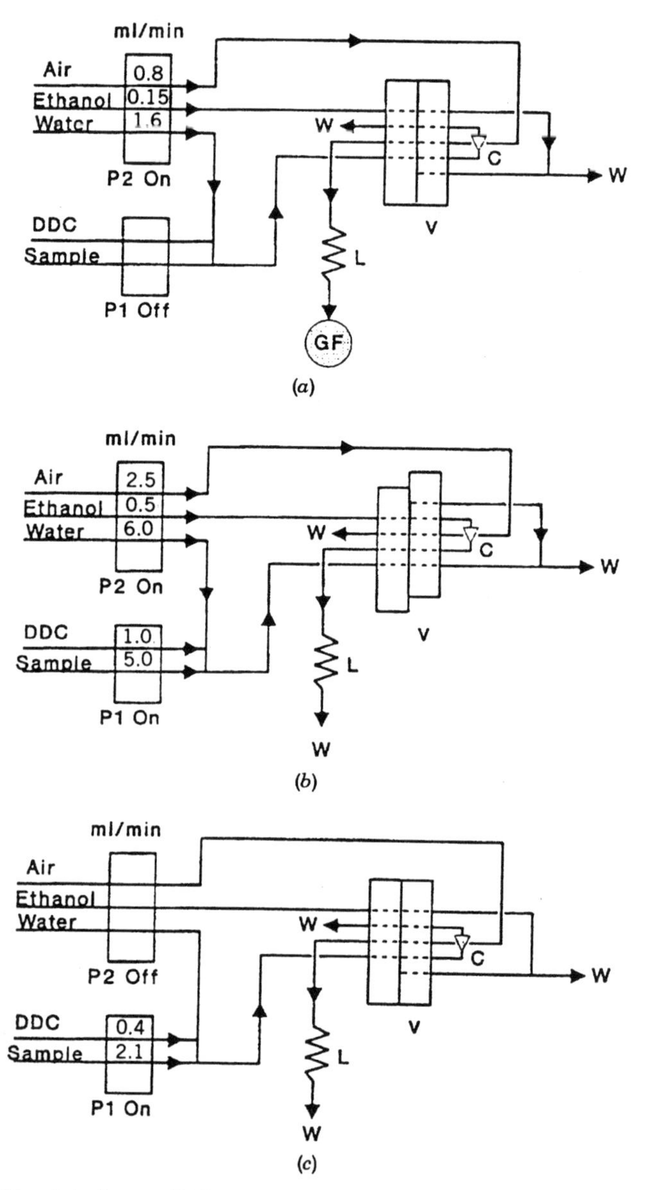

Figure 6.5. Flow injection manifold and operation sequence of automated preconcentration of lead as complexed with diethyldithiocarbamate on line and separated using a 15-μL C_{18} microcolumn: (*a*) dispense eluate into furnace; (*b*) elution of residual analyte and sample change; (*c*) sample loading; (*d*) column rinsing and evacuation of eluate collector; and (*e*) sample elution/eluate collection. P_1, P_2 = peristaltic pumps; C = microcolumn; V = multifunction valve; L = eluate collector (75 μL, 0.35 mm i.d.); W = waste; and GF = graphite furnace. With permission from (195).

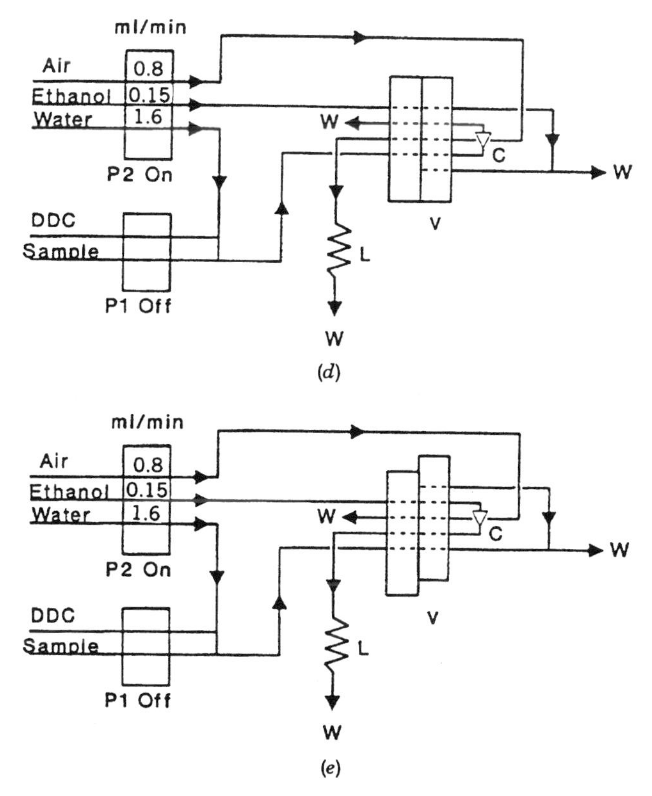

Figure 6.5. *(continued)*

by the application of an appropriate voltage. The authors reported it was necessary to repeat the elution step three times to quantitatively collect the analyte. Each elution was measured separately by GFAAS, and the measurements were added to evaluate each sample.

Flow injection provides a convenient method for automated sample preconcentration, with typical enrichment factors of 20 to 50. We expect this technique to replace batch methods of preconcentration.

Flow injection has also been used with GFAAS as an interface for sample introduction into the furnace, and some applications are listed in Table 6.14. Three groups have developed FI techniques involving aerosol deposition for sample introduction. These approaches were shown to have advantages compared to commercially available (Thermo Jarrell Ash, Franklin, MA) aerosol deposition modules (ADMs) (92) using pneumatic nebulization, which only introduce a small fraction of sample in the furnace.

Table 6.12 Operation Protocol for FI Automated On-line Preconcentration Procedure for GFAAS[a]

Step	Duration, s	Pump	Function	Medium Pumped	Flow rate, mL/min
a	10	1	Dispense eluate from previous preconcentration into furnace	Air	0.8
b	20	1	Elution of residual analyte from previous preconcentration	Ethanol	0.5
		2	Sample change	Sample	5.0
c	60	2	Sample loading on column	Sample	2.1
				DDC solution	0.4
d	30	1	Column rinsing	Water	1.6
		1	Evacuation of eluate collector	Air	0.8
e	30	1	Sample elution and eluate collection	Ethanol	0.15

[a] Lead was complexed with diethyldithiocarbamate (DDC) on line and separated using a 15-μL C_{18} microcolumn (195).

Bank et al. (93,94) described an FI thermospray deposition system. Samples were forced by a high-pressure liquid chromatography (HPLC) pump through a silica capillary into a heated steel tube (300°C), which vaporized the sample. The aerosol was deposited on to a heated graphite tube (optimum sensitivity at 120°C) and water was removed by a vacuum system.

Berndt and Schaldach (95) and Lanza et al. (96) evaluated an FI hydraulic high-pressure nebulization (HHPN) sample introduction system available from KNAUER GmbH (Berlin, Germany). An HPLC pump was used to force a liquid sample through a nozzle to generate a fine aerosol that was collected on a graphite tube heated to 140°C. The efficiency of aerosol transport was significantly improved compared to the commercial ADMs. Lanza et al. (96) employed the HHPN system to determine several elements in synthetic snow and ice. The absorbance was directly proportional to the analyte concentration times the deposition time down to picograms per milliliter concentration levels. For example, a 15-min deposition time was used to introduce 3 pg/mL of cadmium with an RSD of 3 to 5%. High-pressure aerosol deposition systems appear to have some advantages for sample introduction. We expect future applications of these systems for practical analysis.

One interesting application of FI involves its coupling with in situ trapping of volatile hydrides in a graphite tube (97–100). The *hydride generation* (HG)

Table 6.13 Flow Injection Preconcentration/Separation Procedures Coupled with GFAAS

Sample(s)	Analyte(s)	Flow Injection Procedure	References
—	Pd	Commercial FI system with C_{18} column for automated on-line column preconcentration; *N*,*N*-diethyl-*N*′-benzoylthiourea was added to column to complex palladium; impurities were removed with 6.5% nitric acid; palladium was eluted with ethanol; enhancement factors of 40–200	196
Geothermal fluids	Mn	Laboratory-constructed FI system with anion exchange column for preconcentration; samples were transported to the column by water; analyte was eluted with 0.002*M* nitric acid	197
Water	Pb	Commercial FI system with a knotted reactor for automated, on-line preconcentration; diethylthiocarbamate was used to complex lead; elution by ethanol; enhancement factors of 20–140	198
Seawater	Cd, Co, Cu, Ni, Pb	Commercial FI system with column for automated on-line preconcentration; sodium diethyldithiocarbamate was used to complex lead; the chelate was retained on a C_{18} column; enhancement factors of 20–40	83, 195, 199
Wastewater	Pb	Laboratory-constructed FI system with Chelex-100 column for automated on-line preconcentration; enrichment factor of 64	200
Chemical reagents	Pb	Commercial FI system with column for automated on-line preconcentration; the column was a macrocycle immobilized silica gel sorbent; enhancement factor of 77	84
Seawater	Cd, Cu, Fe, Mn, Ni, Pb, Zn	Laboratory-constructed FI system with silica-immobilized 8-hydroxyquinoline; enhancement factor up to 250	201
—	Cd, Co, Cu, Fe, Ni, Pb	Two-step extraction of metal dithiocarbamates using membrane phase separators; final extract collected in a 23-µL sample loop and delivered to the tube; enrichment factors of 50–100	86, 87

(continued)

Table 6.13 ***(continued)***

Sample(s)	Analyte(s)	Flow Injection Procedure	References
Water, sodium sulfate, sodium acetate	Mn	Laboratory-constructed FI system with on-line preconcentration of analyte using a flow-through electrochemical cell	91
Sodium chloride	Pb	Laboratory-constructed FI system with preconcentration on methacrylate gel with immobilized 8-quinolinol	202
Seawater	Cd, Cu	Laboratory-constructed FI system with preconcentration of analyte as the diethyldithiocarbamate on a C_{18} column; enhancement factors of 10–15	203
Tap water	Be	Off-line laboratory-constructed FI system with preconcentration on polyethylene powder for Be-Chrome Azurol S complex; elution with ethanol in reverse direction of preconcentration; comparison of results of a FI method with fluorimetric detection	204
Biological and environmental SRMs	Cd, Cu, Pb	Commercial flow injection system with C_{18} column for automated on-line preconcentration; analytes were complexed with ammonium diethyldithiophosphate with citrate as masking agent; ethanol employed as eluent	205
—	Al, Cd, Cu	Laboratory-constructed automated FI system that included four parallel supported liquid membrane devices; enrichment factor of 10–15	90
Blood	Cd, Ni	Commercial FI system with on-line coprecipitation with an iron(II)–hexamethylenedithiocarbamate complex on the walls of a knotted reactor; the precipitate was dissolved in methyl isobutylketone before sample introduction; enrichment factors of 8–16	88
Seawater	Cu, Pb	Laboratory-constructed FI system with on-line coprecipitation with cobalt–ammonium pyrrolidine dithiocarbamate; precipitate was collected on a PTFE membrane; precipitate was dissolved with methyl isobutyl ketone; enrichment factor of 26	89

Table 6.14 FI Sample Introduction Systems for GFAAS

Sample(s)	Analyte(s)	Flow Injection Procedure	References
—	Ag, Al, As, Au, Cd, Co, Mn, Pb, Ru, V	Laboratory-constructed FI system with thermospray sample deposition; optimized conditions included flow rate of 0.7 mL/min; thermospray temperature of 300°C, and GF deposition temperature of 120°C	93, 94
—	Au, Co, Cu, Pb	Commercial aerosol deposition module (ADM) was modified to act as a hydraulic high-pressure nebulization system (HHPN) by the use of an HPLC pump; a C_{18} column was used for preconcentration of the analytes; enhancement factors up to 100	95
—	As	Laboratory-constructed FI system employed with in situ trapping of volatile hydride into a graphite tube pretreated with palladium (4 μg)	206
Various SRMs	Sn	Commercial FI system employed with in situ trapping of volatile hydride into a graphite tube pretreated with palladium (18 μg)	104
Tap water, hair, serum, geological SRMs	Sn	Commercial FI system employed with column preconcentration and in situ trapping of volatile hydride into a graphite tube pretreated with palladium (10 μg); samples were digested off-line; 2*M* HCl added to samples to form a chlorostannate complex that was retained on a strongly basic anion exchange column; the complex was dissociated by a flow of deionized water into the hydride generation system	85

(*continued*)

Table 6.14 *(continued)*

Sample(s)	Analyte(s)	Flow Injection Procedure	References
—	As, Bi, Se	Commercial FI system employed with in situ trapping of volatile hydride into a graphite tube pretreated with palladium (50 μg) and iridium (50 μg); a single manual injection of the trapping reagents was reported to allow 300 trapping and atomization cycles	207
—	Hg	Commercial FI system employed with in situ trapping of analyte on a gold platinum gauze inserted in a graphite tube	208
Garlic, water, ginseng, geological SRMs	Ge	Commercial FI system employed with in situ trapping of volatile hydride into a graphite tube pretreated with palladium (10 μg); samples were digested off-line; samples merged with borohydride to produce germane that was transferred to a graphite tube maintained at 400°C	105
Water	Sb	Commercial FI system employed with in situ trapping of volatile hydride into a graphite tube	209

technique involves the conversion of the analyte to a volatile hydride with a chemical reagent (usually sodium borohydride), which is then swept into an atom cell (generally a heated quartz tube) where the molecule dissociates in gaseous atoms. Elements that form volatile hydrides include antimony, arsenic, bismuth, germanium, lead, selenium, tellurium, and tin. Other volatile molecules have been used for sample introduction by similar procedures, including chlorides, fluorides, β-diketonates, and dithiocarbamates. In addition, aqueous mercury may be reduced to the metal, which is volatile enough to be determined in a quartz tube maintained at room temperature (cold vapor mercury determination). Several commercial accessories employ quartz tubes that are heated in a flame or electrically as the atomization source with detection by atomic absorption, as described in reviews (101,102) and a book (103). Disadvantages of these conventional HG procedures include dilution of the analyte by carrier gas and hydrogen and low atomization efficiency in quartz tubes due to their relatively low maximum temperatures.

The in situ trapping technique involves flow of hydride into a heated graphite tube that serves to decompose the hydride and condense the analyte on the tube. Recently HG-GFAAS has been reviewed by Matusiewicz and Sturgeon (97). The addition of palladium or platinum group elements induces deposition at relatively low temperatures (200°C) and serves as the chemical modifier in the graphite tube.

Tao and Fang (85) employed an FI-HG-GFAAS system for the determination of tin in hair, tap water, serum, and geological SRMs. Preconcentration of the tin was performed by conversion of the analyte to a chlorostannate complex that adhered to a strongly basic anion exchange column. The tin was eluted from the column by deionized water into the HG system. The tin hydride was collected on a graphite tube coated with palladium.

In general, absolute detection limits are degraded by HG-GFAAS compared to conventional GFAAS (97) since the efficiency of the HG procedure is not 100%. The use of flow injection with HG-GFAAS provides a convenient approach to automate sample introduction procedures (104) and a reduction in interferences compared to a batch system. For example, Tao and Fang (105) demonstrated that an FI-HG-GFAAS system could tolerate an order of magnitude higher nickel concentration when determining germanium. However, in general these methods are relatively difficult to implement, and the sensitivity for most elements is comparable to that obtained by conventional GFAAS. We consequently do not recommend these procedures for routine analysis.

6.4.4 Other Preconcentration/Separation Methods

Several methods of preconcentration are available that have been used for a limited number of applications, which include liquid membrane devices (90), electrochemical cells (91), coprecipitation (88,89), and the use of organisms for preconcentration (106–109). Examples of the first three methods were given in the section on flow injection (Section 6.4.3). Some applications of the use of biological organisms for preconcentration are summarized in Table 6.15.

Robles and Aller (106) described the use of two species of bacteria to preconcentrate beryllium for GFAAS determination. Beryllium was retained on the outer wall at pH 6 to 9 for *Escherichia coli* and pH 3 to 10 for *Pseudomonas putida*. The bacteria were centrifuged to form a pellet that was introduced into a graphite tube as a slurry containing 3.5*M* nitric acid. A preconcentration factor of 6 was reported for both bacterial species.

Majidi and Holcombe (108,109) preconcentrated cadmium in water samples using an alga, *Stichococcus bacillaris*. Freeze-dried alga (10 mg) were mixed with a sample solution, centrifuged, suspended in solution, and introduced into a graphite tube as a slurry. Good accuracy was reported for this method for a

Table 6.15 Use of Organisms for Preconcentration/Separation Coupled to GFAAS

Sample(s)	Analyte(s)	Preconcentration/Separation Procedure	References
Coal flyash SRM	Be	Preconcentration on outer wall of bacteria, *Escherichia coli* and *Pseudomonas putida*; retained by former at pH 6–9 and latter at 3–10; the bacterial mass was dispersed in 3.5*M* nitric acid and introduced as a slurry	106
Tap water	Pb	Preconcentration on chitosan; slurry of chitosan was mixed with sample solution at pH 4–10; the solid was separated by centrifugation; the solid was dissolved in nitric acid	107
Riverine water; seawater	Cd	Preconcentration on 10 mg freeze-dried alga, *Stichococcus bacillaris*, separation by centrifugation; the alga pellet is washed with a pH 7.8 phosphate buffer, suspended as a slurry, and injected into a graphite tube as a slurry in 0.5% nitric acid and phosphate buffer	108, 109

riverine water sample, but preconcentration was not possible in seawater due to the high salinity of these solutions.

In general, the enrichment factors obtained by the bioorganisms is relatively low, and it is often necessary to culture these organisms. In addition, these procedures would appear to be difficult to automate. In general, we would not recommend these procedures for routine analysis; well-established extraction methods are easier to implement and give higher enrichment factors.

6.5 METAL SPECIATION

Metal speciation is the quantitative determination of each of the chemical forms of a metal present in a sample. Considerable interest has developed in speciation over the past 20 years because the toxicity and mobility of metals in the environment and organisms is dependent on their chemical form. Metal compounds may be classified as inorganic, complexed ions, or organometallic. A variation in the toxicity of different oxidation states exists for some metals. For example, chromium(VI) is considerably more toxic than chromium(III). In general, the organometallic compounds are more toxic than inorganic compounds because the former have greater permeability through biomembranes and may accumulate in fatty tissues. Mercury is an example of this type of element, where alkylated mercury compounds (e.g., methyl mercury) are more toxic than inorganic mercury (although these species are also regarded as toxic). Tin compounds (e.g., tributyltin) have been of interest because of their use as algicides, fungicides, and molluscicides. These compounds may accumulate to toxic levels in shellfish and fish, although inorganic tin is an essential trace element. Arsenic is an exception to the general rule because some organometallic forms, such as arsenobetaine, arsenocholine, and some arsenosugars, are relatively nontoxic, but inorganic arsenic(III) (arsenite) and arsenic(V) (arsenate) are toxic.

A considerable body of literature is available on metal speciation. Volumes have been edited by Kramer and Allen (110) and Harrison and Rapsomanikis (111). The latter contains a chapter on speciation by atomic absorption (112), as well as chapters on individual elements such as tin, germanium, lead, arsenic, antimony, mercury, and selenium. Cornelius and De Kimpe (113) reviewed elemental speciation in biological fluids, and Irgolic (114) discussed procedures to do speciation of environmental samples. The volume edited by Prichard and co-workers (115) includes a chapter on metal speciation. Quevauviller (116) reviewed projects conducted by a testing program conducted by the Community Bureau of Reference of the European Union to assess the accuracy and potential sources of error in metal speciation. Here we discuss some general aspects of speciation with an emphasis on some recent GFAAS applications.

The various chemical forms of a metal must be separated by a method that does not change the chemical structure of the analytes (116) prior to detection by GFAAS or another method. Perhaps the most commonly used separation technique is extraction, either with acids or organic solvents. It is necessary to verify the recovery of the procedure by measurement of the extraction recovery for each analyte. This procedure involves spiking a sample with each analyte and measuring the concentration after extraction. An alternative procedure is derivatization of analytes to achieve preconcentration of the analytes. For example, hydride generation can be employed to preconcentrate hydride-forming elements (117). Alternatively, Grignard reactions may be employed to induce pentylation of alkyllead and alkyltin species and produce compounds that can be separated easily by gas chromatography. Derivatization methods may lead to errors because of incompleteness of reactions (e.g., arsenobetaine is not converted to a volatile hydride by sodium borohydride) and probably should be avoided when possible.

Some examples of metal speciation with GFAAS detection are given in Table 6.16. Separation of the analytes has been achieved by a number of procedures, including gas chromatography (118), liquid chromatography (119), extraction (120), and coprecipitation (121,122). Ferri and co-workers (119) described a procedure to determine tin and triorganotin compounds (triphenyl tin, dibutyltin, tributyltin, and monobutyl tin) in seawater using a silica gel column for separation with GFAAS detection. The organic compounds were preconcentrated on graphitized carbon black and eluted from the silica column by hexane/ethyl acetate mixtures. Tin(IV) was determined by formation of an ammonium pyrrolidinedithiocarbamate complex and extraction into methylene chloride. Good recovery ($\geq 95\%$) and precision ($\leq 5\%$) were reported at nanogram per liter concentration levels.

A disadvantage of GFAAS with many conventional separation procedures is its incompatibility with a flowing system. These problems may be alleviated by doing the separations in parallel by flow injection methods (Section 6.4.3). Several on-line sorption speciation methods have been reported (123–125) as discussed in Section 6.4.3. Le and co-workers (117) reported speciation of arsenic using conventional hydride generation and in situ trapping of the hydride in a graphite tube. Arsenic(V) compounds were shown to have low reactivity toward sodium borohydride, resulting in low recoveries. The addition of cysteine was shown to reduce arsenic(V) to arsenic(III) and minimize these interferences.

Bendicho (126) coupled an HPLC system to a GFAAS system using an automated thermospray interface. Sample introduction was performed by periodic deposition of eluent into a preheated custom-designed graphite platform. This system was employed for the speciation of arsenate, arsenite, and dimethylarsinate using reverse-phase isocratic elution with an RSD of 6%.

Table 6.16 Applications of Speciation by GFAAS

Sample(s)	Analyte(s)	Speciation Procedure	References
Seawater	As(III), As	As(III): Commercial flow injection system with column for automated on-line preconcentration; sodium diethyldithiocarbamate was used to complex As(III); chelate was retained on a C_{18} column; enrichment factor of 20–40. Total As: As(V) was reduced to As(III) on line.	123
Water	Sb(III)	Commercial flow injection system with a knotted reactor for automated, on-line preconcentration; pyrolidine dithiocarbamate was used to complex Sb(III); elution of complex from inner walls of reactor with ethanol; enhancement factor of 30	124
Urine	As(III), As	Laboratory-constructed system coupling flow injection with in situ trapping in a graphite furnace; HCl and $NaBH_4$ were used to generate the hydrides, cysteine was shown reduce interferences, presumably by conversion of As(V) to As(III)	125
Water	Cr(VI), Cr	Commercial flow injection system with column for automated on-line preconcentration; sodium diethyldithiocarbamate was used to complex Cr(VI); total Cr determined by oxidation of Cr(III) to Cr(VI); enhancement factor of 12	126
Seawater	Cr(III), Cr(VI)	Cr(III) determined by coprecipitation with gallium hydroxide, centrifuged, and dissolved in concentrated nitric acid; for total Cr, Cr(VI) converted to Cr(III) with hydroxylammonium chloride; preconcentration performed by the above procedure	121
Animal feeds	Cr, Cr(VI)	Total Cr determined by a wet digestion method; Cr(III) was precipitated with NaOH; addition of 1*M* NH_4NO_3, centrifugation, and measurement of Cr(VI) in the supernatant	122

(*continued*)

Table 6.16 ***(continued)***

Sample(s)	Analyte(s)	Speciation Procedure	References
Wood preservatives	Dibutyltin, bis(tributyltin)oxide, tributyltin	HPLC with a cation exchange column interfaced to a continuously heated GFAAS system	210
Sediments	As, As(III)	As(III): sediments (0.1 g) were treated with water pH 4.63 buffer and analyte was extracted by sodium diethyldithiocarbamate into methyl isobutyl ketone; the organic phase was introduced into a graphite tube. As: Sediments were ground to a particle size $<250\,\mu m$; agitated with zirconia beads in water; treated with Triton X-100; transferred to an autosampler cup and stirred magnetically before sample introduction	120
Seawater	Sn, Sn(IV), triphenyltin, tributyltin	Sn(IV): Sn(IV) was extracted converted to the ammonium pyrrolidinedithiocarbamate complex and extracted into methylene chloride. Organotin compounds: graphitized carbon black was used to preconcentrate the compounds; analytes were loaded with deionized water and eluted with methanol/methylene chloride (4+1); solvent was evaporated; compounds were separated on a silica gel column (200 mg); tributyltin was eluted with hexane/ethyl acetate (2+1) and triphenyltin was eluted with ethyl acetate	119
Air	Alkyllead compounds	Air samples were bubbled through an aqueous iodine monochloride solution to trap the analytes; analytes were extracted into *n*-hexane and propylated or butylated and determined by gas chromatography–GFAAS	118

Air	Alkyllead compounds	Alkyllead compounds were adsorbed on to activated carbon; extracted with hot nitric acid and determined by GFAAS; particulate inorganic lead was collected on a membrane filter in front of absorbent	211
—	As(III), As(V), dimethylarsinate	HPLC system with C_{18} column was used for separation of the compounds; automated flow injection thermospray interface was used to introduce samples into a graphite tube with custom designed platforms for thermospray deposition	126
Hemin	Fe	Selective volatilization from a two-step furnace gave two peaks, one due to heme iron and the other due to nonheme iron; no quantitative data were reported	127
Serum	Al-transferrin (Al-Tf), Al-desferrioxamine (DFO), Al-citrate, Si-Tf, Si-DFO, Si-citrate	Separation by HPLC with an anion-exchange column; results were compared to previous studies using complementary techniques	212

Smith and Harnly (127) described a selective volatilization approach for the determination of iron in a porphyrin complex (heme iron) and nonheme iron. They obtained two temporally separated iron peaks for a hemin standard using a two-step furnace. The first peak, which vaporized at 300°C, was attributed to heme iron, and the second, which vaporized at 2000°C, to nonheme iron. The iron concentrations were not quantified because the purity of the standard was unknown and complete dissolution of the hemin was not achieved. Although clearly more work needs to be done in this area, quantitative direct speciation by GFAAS without any separation steps would certainly reduce analysis time and complexity.

In conclusion, although traditional methods of speciation may be difficult to interface with GFAAS, flow injection provides a convenient way to determine various forms of elements in an on-line, automated fashion. We expect a number of new methods to be developed in this area.

REFERENCES

1. M. Hoenig and A.-M. Kersabiec, *Spectrochim. Acta, Part B*, **51B**, 1297 (1996).
2. K. D. Ohls, *Spectroscopy*, **12**(1), 19 (1997).
3. R.F. Browner, *Anal. Chem.*, **56**, 786A (1984).
4. S. J. Haswell, Ed., *Atomic Absorption Spectrometry: Theory, Design, and Applications*, Analytical Spectroscopy Library Ed., Vol. 5, Elsevier, Amsterdam, 1991.
5. A. Leblanc, *J. Anal. Atom. Spectrom.*, **11**, 1093 (1996).
6. B. T. Kildahl and W. Lund, *Fresenius J. Anal. Chem.*, **354**, 93 (1996).
7. H. van der Jagt and P. J. Stuyfzand, *Fresenius J. Anal. Chem.*, **354**, 32 (1996).
8. A. Deval and J. Sneddon, *Microchem. J.*, **52**, 96 (1995).
9. K. Tompuri and J. Tummavuori, *Anal. Comm.*, **33**, 283 (1996).
10. J. Marshall, "Applications of Atomic Absorption Spectrometry in the Petroleum Industry," in S. J. Haswell, Ed., *Atomic Absorption Spectrometry: Theory, Design, and Application*, Elsevier, Amsterdam, 1991.
11. *NIST Standard Reference Materials Catalog*, Gaithersburg, MD, 1996.
12. C. S. Saba, W. E. Rhine, and K. J. Eisentraut, *Appl. Spectrosc.*, **39**, 1985 (1985).
13. J. Sneddon, *Talanta*, **30**, 631 (1983).
14. J. P. Lodge, Ed., *Methods of Air Sampling and Analysis*, Lewis, Chelsea, MI, 1989.
15. J. C. Septon, "Analysis of Airborne Particles in Workplace Atmospheres," in S. J. Haswell, Ed., *Atomic Absorption Spectrometry: Theory, Design, and Application*, Elsevier, Amsterdam, 1991.
16. U. Telgheder and V. A. Khvostikov, *J. Anal. Atom. Spectrom.*, **12**, 1 (1997).
17. C.-F. Wang, J.-Y. Yang, and C.-H. Ke, *Anal. Chim. Acta*, **320**, 207 (1996).

18. Y.-I. Lee, M. V. Smith, S. Indurthy, A. Deval, and J. Sneddon, *Spectrochim. Acta, Part B*, **51B**, 109 (1996).
19. J. Sneddon, M. V. Smith, S. Indurthy, and Y.-I. Lee, *Spectroscopy*, **10**(1), 26 (1995).
20. Z. Liang, G.-T. Wei, R. L. Irwin, A. P. Walton, R. G. Michel, and J. Sneddon, *Anal. Chem.*, **62**, 1452 (1990).
21. J. Sneddon, *Appl. Spectrosc.*, **44**, 1562 (1990).
22. J. Sneddon, *Anal. Chim. Acta*, **245**, 203 (1991).
23. G. Torsi, P. Reschligian, F. Fagioli, and C. Locatelli, *Spectrochim. Acta, Part B*, **48B**, 681 (1993).
24. G. Torsi, *Spectrochim. Acta, Part B*, **50B**, 707 (1995).
25. G. Torsi, P. Reschiglian, M. T. Lippolis, and A. Toschi, *Microchem. J.*, **53**, 437 (1996).
26. C. Locatelli, P. Reschiglian, G. Torsi, F. Fagioli, N. Rossi, and D. Melucci, *Appl. Spectrosc.*, **50**, 1585 (1996).
27. P. Bermejo-Barrera, A. Moreda-Pineiro, J. Moreda-Pineiro, and A. Bermejo-Barrera, *J. Anal. Atom. Spectrom.*, **10**, 1011 (1995).
28. M. Hoenig and R. de Borger, *Spectrochim. Acta, Part B*, **38B**, 873 (1983).
29. C. Vandecasteele and C. B. Block, *Modern Methods for Trace Element Determination*, Wiley, New York, 1993.
30. S. B. McLaughlin, M. G. Tjoelker, and W. K. Roy, *Can. J. For. Res.*, **23**, 380 (1993).
31. A. L. Page, R. H. Miller, and D. R. Keeney, Eds., *Methods of Soil Analysis, Part 2: Chemical and Microbiological Properties*, Vol. 2, American Society of Agronomy and Soil Science Society of America, Madison, WI, 1982.
32. H. M. Kingston and L. B. Jassie, *Introduction to Microwave Sample Preparation*, American Chemical Society, Washington, DC, 1988.
33. H. Matusiewicz and R. E. Sturgeon, *Prog. Anal. Spectrosc.*, **12**, 21 (1989).
34. H. M. Kingston and P. J. Walter, *Spectroscopy*, **7**(9), 20 (1992).
35. M. A. Z. Arruda, M. Gallego, and M. Valcárcel, *J. Anal. Atom. Spectrom.*, **10**, 501 (1995).
36. M. A. Z. Arruda, M. Gallego, and M. Valcárcel, *J. Anal. Atom. Spectrom.*, **11**, 169 (1996).
37. R. E. Sturgeon, S. N. Willie, B. A. Methven, J. W. H. Lam, and H. Matusiewicz, *J. Anal. Atom. Spectrom.*, **10**, 981 (1995).
38. M. Burguera, J. L. Burguera, C. Rondón, C. Rivas, P. Carrero, M. Gallignani, and M. R. Brunetto, *J. Anal. Atom. Spectrom.*, **10**, 343 (1995).
39. B. D. Zehr and T. J. Zehr, *Spectroscopy*, **6**(4), 44 (1991).
40. M. Grognard and M. Piolon, *Atom. Spectrosc.*, **6**, 142 (1985).
41. D. L. McCurdy, A. E. Weber, S. K. Hughes, and R. C. Fry, in J. Sneddon, Ed., *Sample Introduction in Atomic Spectroscopy*, Chapter 3, Elsevier, Amsterdam, 1990.
42. N. J. Miller-Ihli, *Anal. Chem.*, **64**, 964A (1992).
43. F. J. Langmyhr and G. Wibetoe, *Prog. Anal. Atom. Spectrosc.*, **8**, 193 (1985).

44. C. Bendicho and M. T. C. de Loos-Vollebregt, *J. Anal. Atom. Spectrom.*, **6**, 353 (1991).
45. Z. A. de Benzo, M. Velosa, C. Ceccarelli, M. de la Guardia, and A. Salvador, *Fresenius J. Anal. Chem.*, **339**, 235 (1991).
46. N. J. Miller-Ihli, *J. Anal. Atom. Spectrom.*, **4**, 295 (1989).
47. G. R. Carnrick, G. Daley, and A. Fotinopoulos, *Atom. Spectrosc.*, **10**, 170 (1989).
48. N. J. Miller-Ihli, *J. Anal. Atom. Spectrom.*, **3**, 73 (1988).
49. Y. Tan, W. D. Marshall, and J.-S. Blais, *Analyst*, **121**, 483 (1996).
50. Y. Tan, J.-S. Blais, and W. D. Marshall, *Analyst*, **121**, 1419 (1996).
51. U. Schäffer and V. Krivan, *J. Anal. Atom. Spectrom.*, **11**, 1119 (1996).
52. N. J. Miller-Ihli, *Spectrochim. Acta, Part B*, **50B**, 477 (1995).
53. N. J. Miller-Ihli, *J. Anal. Atom. Spectrom.*, **12**, 205 (1997).
54. N. J. Miller-Ihli, *J. Anal. Atom. Spectrom.*, **9**, 1129 (1994).
55. G. R. Carnrick, B. K. Lumas, and W. B. Barnett, *J. Anal. Atom. Spectrom.*, **1**, 443 (1986).
56. G. R. Dulude and J. J. Sotera, *Spectrochim. Acta, Part B*, **39B**, 511 (1984).
57. U. Kurfürst, M. Kempeneer, M. Stoeppler, and O. Schuierer, *Fresenius J. Anal. Chem.*, **337**, 248 (1990).
58. G. van Dalen, *J. Anal. Atom. Spectrom.*, **11**, 1087 (1996).
59. E. Lücker and O. Schuierer, *Spectrochim. Acta, Part B*, **51B**, 201 (1996).
60. U. Kurfürst, A. Rehnert, and H. Muntau, *Spectrochim. Acta, Part B*, **51B**, 229 (1996).
61. M. Berglund and D. C. Baxter, *Spectrochim. Acta, Part B*, **47B**, E1567 (1992).
62. R. L. Irwin, A. Mikkelsen, R. G. Michel, J. P. Dougherty, and F. P. Preli, *Spectrochim. Acta, Part B*, **45B**, 903 (1990).
63. R. E. Russo, *Appl. Spectrosc.*, **49**, 14A (1995).
64. P. M. Outridge, *Spectroscopy*, **11**(4), 21 (1995).
65. T. L. Thiem, *Am. Lab.*, **26**(2), 48CCC (1994).
66. E. R. Denoyer, K. J. Fredeen, and J. W. Hager, *Anal. Chem.*, **63**, 445A (1991).
67. D. Günther, H. P. Longerich, L. Forsythe, and S. E. Jackson, *Am. Lab.*, **27**(6), 24 (1995).
68. J. X. Zhou, S.-J. J. Tsai, X. Hou, K. X. Yang, and R. G. Michel, *Microchem. J.*, **54**, 111 (1996).
69. R. Wennrich and K. Dittrich, *Spectrochim. Acta, Part B*, **42B**, 995 (1987).
70. M. S. Cresser, *Solvent Extraction in Flame Spectroscopic Analysis*, Butterworths, London, 1978.
71. M. Blankley, A. Henson, and K. C. Thompson, "Water, Sewage, and Effluents," in S. J. Haswell, Ed., *Atomic Absorption Spectrometry: Theory, Design, and Application*, Elsevier, Amsterdam, 1991.
72. J. Komárek and L. Sommer, *Talanta*, **29**, 159 (1982).

73. H. Haraguchi and T. Akagi, "Application of Atomic Absorption Spectrometry to Marine Analysis," in S. J. Haswell, Ed., *Atomic Absorption Spectrometry: Theory, Design, and Application*, Elsevier, Amsterdam, 1991.
74. L. Ebdon, S. Hill, and R. W. Ward, *Analyst*, **111**, 1113 (1986).
75. L. C. Ebdon, S. Hill, and R. W. Ward, *Analyst*, **112**, 1 (1987).
76. J. Tyson, *Analyst*, **110**, 419 (1985).
77. J. F. Tyson, *Spectrochim. Acta Rev.*, **14**, 169 (1991).
78. J. Tyson, *Spectroscopy*, **7**(3), 14 (1992).
79. Z. Fang, *Spectrochim. Acta Rev.*, **14**, 235 (1991).
80. Z. Fang, *Flow Injection Separation and Preconcentration*, VCH, Weinheim, Germany, 1993.
81. Z. Fang, *Flow Injection Atomic Absorption Spectrometry*, Wiley, New York, 1995.
82. Z. Fang, S. Xu, and G. Tao, *J. Anal. Atom. Spectrom.*, **11**, 1 (1996).
83. M. Sperling, X. Yin, and B. Welz, *J. Anal. Atom. Spectrom.*, **6**, 615 (1991).
84. M. Sperling, X.-P. Yan, and B. Welz, *Spectrochim. Acta, Part B*, **51B**, 1875 (1996).
85. G. Tao and Z. Fang, *Talanta*, **42**, 375 (1995).
86. K. Bäckström and L.-G. Danielsson, *Anal. Chem.*, **60**, 1354 (1988).
87. K. Bäckström and L.-G. Danielsson, *Anal. Chim. Acta*, **232**, 301 (1990).
88. Z. Fang and L. Dong, *J. Anal. Atom. Spectrom.*, **7**, 439 (1992).
89. Z. Zhuang, X. Wang, P. Yang, C. Yang, and B. Huang, *Can. J. Appl. Spectrosc.*, **39**, 101 (1994).
90. F. Malcus, N. K. Djane, L. Mathiasson, and G. Johansson, *Anal. Chim. Acta*, **327**, 295 (1996).
91. E. Beinrohr, M. Rapta, M.-L. Lee, P. Tschöpel, and G. Tölg, *Mikrochim. Acta*, **110**, 1 (1993).
92. J. J. Sotera, L. C. Cristiano, M. K. Conley, and H. L. Kahn, *Anal. Chem.*, **55**, 204 (1983).
93. P. C. Bank, M. T. C. de Loos-Vollebregt, and L. de Galan, *Spectrochim. Acta, Part B*, **43B**, 983 (1988).
94. P. C. Bank, M. T. C. de Loos-Vollebregt, and L. de Galan, *Spectrochim. Acta, Part B*, **44B**, 571 (1989).
95. H. Berndt and G. Schaldach, *J. Anal. Atom. Spectrom.*, **9**, 39 (1994).
96. F. Lanza, A. Ceccarini, and P. Papoff, *Spectrochim. Acta, Part B*, **52B**, 113 (1997).
97. H. Matusiewicz and R. E. Sturgeon, *Spectrochim. Acta, Part B*, **51B**, 377 (1996).
98. D. S. Lee, *Anal. Chem.*, **54**, 1682 (1982).
99. R. E. Sturgeon, S. N. Willie, and S. S. Berman, *Anal. Chem.*, **57**, 2311 (1985).
100. R. E. Sturgeon, S. N. Willie, G. I. Sproule, and S. S. Berman, *J. Anal. Atom. Spectrom.*, **2**, 719 (1987).
101. T. Nakahara, *Prog. Anal. Atom. Spectrom.*, **6**, 163 (1983).
102. A. D. Campbell, *Pure Appl. Chem.*, **64**, 227 (1992).

103. J. Dedina and D. L. Tsalev, *Hydride Generation Atomic Absorption Spectrometry*, Chemical Analysis. A Series of Monographs on Analytical Chemistry and Its Application, J. D. Winefordner, Ed., Vol. 130, Wiley, New York, 1995.

104. Z. Li, S. McIntosh, G. R. Carnrick, and W. Slavin, *Spectrochim. Acta, Part B*, **47B**, 701 (1992).

105. G. Tao and Z. Fang, *J. Anal. Atom. Spectrom.*, **8**, 577 (1993).

106. L. C. Robles and A. J. Aller, *J. Anal. Atom. Spectrom.*, **9**, 871 (1994).

107. Y. Tang, B. Chen, and S. Mo, *Talanta*, **43**, 761 (1996).

108. V. Majidi and J. A. Holcombe, *Spectrochim. Acta, Part B*, **43B**, 1423 (1988).

109. V. Majidi and J. A. Holcombe, *J. Anal. Atom. Spectrom.*, **4**, 439 (1989).

110. J. R. Kramer and H. E. Allen, Eds., *Metal Speciation; Theory, Analysis, and Application*, Lewis, Chelsea, MI, 1988.

111. R. M. Harrison and S. Rapsomanikis, Eds., *Environmental Analysis using Chromatography Interfaced with Atomic Spectroscopy*, Halsted, New York, 1989.

112. C. N. Hewitt, "Atomic Absorption Detectors," in R. M. Harrison and S. Rapsomanikis, Eds., *Environmental Analysis Using Chromatography Interfaced with Atomic Spectroscopy*, Wiley, New York, 1989.

113. R. Cornelius and J. De Kimpe, *J. Anal. Atom. Spectrom.*, **9**, 945 (1994).

114. K. J. Irgolic, *Sci. Total Environ.*, **64**, 61 (1987).

115. E. Prichard, G. M. MacKay, and J. Points, Eds. *Trace Analysis: A Structured Approach to Obtaining Reliable Results*, Royal Society of Chemistry, Cambridge, UK, 1996.

116. P. Quevauviller, *J. Anal. Atom. Spectrom.*, **11**, 1225 (1996).

117. X.-C. Le, W. R. Cullen, and K. J. Reimer, *Anal. Chim. Acta*, **285**, 277 (1994).

118. C. N. Hewitt, R. M. Harrison, and M. Radojevic, *Anal. Chim. Acta*, **188**, 229 (1986).

119. T. Ferri, E. Cardarelli, and B. M. Petronio, *Talanta*, **36**, 513 (1989).

120. P. Bermejo-Barrera, M. C. Barciela-Alonso, M. Ferrón-Novais, and A. Bermejo-Barrera, *J. Anal. Atom. Spectrom.*, **10**, 247 (1995).

121. A. Boughriet, L. Deram, and M. Wartel, *J. Anal. Atom. Spectrom.*, **9**, 1135 (1994).

122. M. E. Soares, M. L. Bastos, and M. A. Ferreira, *J. Anal. Atom. Spectrom.*, **9**, 1269 (1994).

123. M. Sperling, X. Yin, and B. Welz, *Spectrochim. Acta, Part B*, **46B**, 1789 (1991).

124. X.-P. Yan, W. Van Mol, and F. Adams, *Analyst*, **121**, 1061 (1996).

125. M. Sperling, X. Yin, and B. Welz, *Analyst*, **117**, 629 (1992).

126. C. Bendicho, *Anal. Chem.*, **66**, 4375 (1994).

127. C. M. M. Smith and J. M. Harnly, *J. Anal. Atom. Spectrom.*, **11**, 1055 (1996).

128. S.-D. Huang and K.-Y. Shih, *Spectrochim. Acta, Part B*, **48B**, 1451 (1993).

129. H. Chuang and S.-D. Huang, *Spectrochim. Acta, Part B*, **49B**, 283 (1994).

130. E. A.-C. Cimadevilla, K. Wróbel, J. M. M. Gayón, and A. Sanz-Medel, *J. Anal. Atom. Spectrom.*, **9**, 117 (1994).

131. J. M. Pérez Parajón and A. Sanz-Medel, *J. Anal. Atom. Spectrom.*, **9**, 111 (1994).

132. M. Patriarca and G. S. Fell, *J. Anal. Atom. Spectrom.*, **9**, 457 (1994).

133. J. Y. Cabon and A. Le Bihan, *J. Anal. Atom. Spectrom.*, **9**, 477 (1994).

134. E. Knutsen, G. Wibetoe, and I. Martinsen, *J. Anal. Atom. Spectrom.*, **10**, 757 (1995).

135. X. Dong-Qun, G. Gang-Ping, and S. Han-Wen, *J. Anal. Atom. Spectrom.*, **10**, 753 (1995).

136. M. Fukushima, T. Ogata, K. Haraguchi, K. Nakagawa, S. Ito, M. Sumi, and N. Asami, *J. Anal. Atom. Spectrom.*, **10**, 999 (1995).

137. L. M. Briggs-Reid and M. G. Heagler, *Microchem. J.*, **55**, 122 (1997).

138. M. G. Heagler, A. G. Lindow, J. N. Beck, C. S. Jackson, and J. Sneddon, *Microchem. J.*, **53**, 472 (1996).

139. I. Karadjova, *Microchem. J.*, **54**, 144 (1996).

140. J. M. Anzano, E. Perisé, M. A. Belarra, and J. R. Castillo, *Microchem. J.*, **52**, 268 (1995).

141. P. Bermejo-Barrera, A. Moreda-Piñeiro, J. Moreda-Piñeiro, and A. Bermejo-Barrera, *Talanta*, **43**, 1783 (1996).

142. Y. Y. Zong, P. J. Parsons, and W. Slavin, *J. Anal. Atom. Spectrom.*, **11**, 25 (1996).

143. M. A. Z. Arruda, M. J. Quinela, M. Gallego, and M. Valcárcel, *Analyst*, **119**, 1695 (1994).

144. M. A. Z. Arruda, M. Gallego, and M. Valcárcel, *J. Anal. Atom. Spectrom.*, **10**, 55 (1995).

145. D. D. T. Phuong, E. Tatár, I. Varga, G. Záray, E. Cseh, and F. Fodor, *Microchem. J.*, **51**, 145 (1995).

146. P. Anderson, C. M. Davidson, D. Littlejohn, A. M. Ure, C. A. Shand, and M. V. Cheshire, *Anal. Chim. Acta*, **327**, 53 (1996).

147. R. Tahvonen and J. Kumpulainen, *J. Anal. Atom. Spectrom.*, **9**, 1427 (1994).

148. Z. Yuan-fu, Z. Ke, F. Zheng, and W. Yun-Zhou, *J. Anal. Atom. Spectrom.*, **10**, 359 (1995).

149. R. Chakraborty, A. K. Das, M. L. Cervera, and M. de la Guardia, *J. Anal. Atom. Spectrom.*, **10**, 353 (1995).

150. I. Harrison, D. Littlejohn, and G. S. Fell, *J. Anal. Atom. Spectrom.*, **10**, 215 (1995).

151. M. Taddia, A. Bellini, and R. Fornari, *J. Anal. Atom. Spectrom.*, **10**, 433 (1995).

152. J. Srámková, S. Kotrlý, and K. Dolezalová, *J. Anal. Atom. Spectrom.*, **10**, 763 (1995).

153. B. Mile, C. C. Rowlands, and A. V. Jones, *J. Anal. Atom. Spectrom.*, **10**, 785 (1995).

154. S. Arpadjan and A. Alexandrova, *J. Anal. Atom. Spectrom.*, **10**, 799 (1995).

155. M. Yizai, L. Zhikun, W. Xiaohui, W. Jiazhen, and L. Yongquan, *J. Anal. Atom. Spectrom.*, **9**, 679 (1994).

156. H. Zhuoer, *J. Anal. Atom. Spectrom.*, **9**, 11 (1994).

157. R. Saraswati, N. R. Desikan, and T. H. Rao, *J. Anal. Atom. Spectrom.*, **9**, 1289 (1994).

158. E. Wieteska, A. Zióek, and A. Drzewinska, *Anal. Chim. Acta*, **330**, 251 (1996).

159. N. J. Miller-Ihli, *Spectrochim. Acta, Part B*, **44B**, 1221 (1989).

160. N. J. Miller-Ihli, *Fresenius J. Anal. Chem.*, **337**, 271 (1990).

161. B. Doecekal and V. Krivan, *Spectrochim. Acta, Part B*, **48B**, 1645 (1993).

162. D. J. Butcher, R. L. Irwin, J. Takahashi, G. Su, G.-T. Wei, and R. G. Michel, *Appl. Spectrosc.*, **44**, 1521 (1990).

163. C. Bendicho and M. T. C. de Loos-Vollebregt, *Spectrochim. Acta, Part B*, **45B**, 695 (1990).

164. E. D. Byrd and D. J. Butcher, *Spectrosc. Lett.*, **26**, 1613 (1993).

165. R. Dobrowolski, *Spectrochim. Acta, Part B*, **51B**, 221 (1996).

166. R. Dobrowolski and J. Mierzwa, *Analyst*, **121**, 897 (1996).

167. D. Bradshaw and W. Slavin, *Spectrochim. Acta, Part B*, **44B**, 1245 (1989).

168. I. López-García, M. Sánchez-Merlos, and M. Hernández-Córdoba, *Anal. Chim. Acta*, **328**, 19 (1996).

169. I. López-García, M. Sánchez-Merlos, and M. Hernández-Córdoba, *J. Anal. Atom. Spectrom.*, **11**, 1003 (1996).

170. U. Schäffer and V. Krivan, *Spectrochim. Acta, Part B*, **51B**, 1211 (1996).

171. S. Hauptkorn, G. Schneider, and V. Krivan, *J. Anal. Atom. Spectrom.*, **9**, 463 (1994).

172. P. Bermejo-Barrera, C. Barciela-Alonso, M. Aboal-Somoza, and A. Bermejo-Barrera, *J. Anal. Atom. Spectrom.*, **9**, 469 (1994).

173. P. Bermejo-Barrera, M. C. Barciela-Alonso, J. Moreda-Pineiro, C. González-Sixto, and A. Bermejo-Barrera, *Spectrochim. Acta, Part B*, **51B**, 1235 (1996).

174. J. Stupar and F. Dolinsek, *Spectrochim. Acta, Part B*, **51B**, 665 (1996).

175. M. Papaspyrou, L. E. Feinendegen, C. Mohl, and M. J. Schwuger, *J. Anal. Atom. Spectrom.*, **9**, 791 (1994).

176. P. Bermejo-Barrera, A. Moreda-Piñeiro, J. Moreda-Piñeiro, and A. Bermejo-Barrera, *Talanta*, **43**, 1099 (1996).

177. U. Osberghaus, U. Kurfürst, and M. Stoeppler, *Fresenius J. Anal. Chem.*, **322**, 739 (1985).

178. M. W. Hinds, *Spectrochim. Acta, Part B*, **48B**, 435 (1993).

179. M. W. Hinds and V. V. Kogan, *J. Anal. Atom. Spectrom.*, **9**, 451 (1994).

180. G. Schlemmer and B. Welz, *Fresenius J. Anal. Chem.*, **328**, 405 (1987).

181. K.-C. Friese, V. Krivan, and O. Schuierer, *Spectrochim. Acta, Part B*, **51B**, 1223 (1996).

182. H. Docekal and V. Krivan, *Spectrochim. Acta, Part B*, **50B**, 517 (1995).

183. J. M. Anzano, M. P. Martinez-Garbayo, M. A. Belarra, and J. R. Castillo, *J. Anal. Atom. Spectrom.*, **9**, 125 (1994).

184. M. M. Smith, M. A. White, and H. K. Wilson, *J. Anal. Atom. Spectrom.*, **10**, 349 (1995).

185. I. Rivera-Duarte and A. R. Flegal, *Anal. Chim. Acta*, **328**, 13 (1996).

186. V. G. Torgov, M. G. Demidova, T. M. Korda, N. K. Kalish, and R. S. Shulman, *Analyst*, **121**, 489 (1996).

187. M. Lan-Xiang, V. Näntö, L. Song-Ping, M. Walls, A. L. Mäkelä, W. Wang, P. Laihonen, M. Ketola, M. Reinikainen, M. Hämäläinen, E. Häsänen, P. Manninen, U. Wegelius, P. Mäkelä, and L. Zhi-Min, *Microchem. J.*, **52**, 223 (1995).

188. E. Tservosky, S. Arpadjan, and I. Karadjova, *Spectrochim. Acta, Part B*, **47B**, 959 (1992).

189. C.-H. Chung, E. Iwamoto, M. Yamamoto, and Y. Yamamoto, *Spectrochim. Acta, Part B*, **39B**, 459 (1984).

190. B. Magnusson and S. Westerlund, *Anal. Chim. Acta*, **131**, 63 (1981).

191. L. Liang, P. C. D'Haese, L. V. Lamberts, F. L. Van de Vyver, and M. E. De Broe, *Anal. Chem.*, **63**, 423 (1991).

192. R. E. Sturgeon, S. S. Berman, and S. N. Willie, *Talanta*, **29**, 167 (1982).

193. D. Wildhagen, V. Krivan, B. Gercken, and J. Pavel, *J. Anal. Atom. Spectrom.*, **11**, 371 (1996).

194. A. K. Avila and A. J. Curtius, *J. Anal. Atom. Spectrom.*, **9**, 543 (1994).

195. Z. Fang, M. Sperling, and B. Welz, *J. Anal. Atom. Spectrom.*, **5**, 639 (1990).

196. M. Schuster and M. Schwaraer, *Anal. Chim. Acta*, **328**, 1 (1996).

197. J. L. Burguera, M. Burguera, C. Rivas, P. Carrero, M. Gallignani, and M. R. Brunetto, *J. Anal. Atom. Spectrom.*, **10**, 479 (1995).

198. M. Sperling, X.-P. Yan, and B. Welz, *Spectrochim. Acta, Part B*, **51B**, 1891 (1996).

199. M. Sperling, X. Yin, and B. Welz, *J. Anal. Atom. Spectrom.*, **6**, 295 (1991).

200. M. M. Silva, F. J. Krug, P. V. Oliveira, J. A. Nóbrega, B. F. Reis, and D. A. G. Penteado, *Spectrochim. Acta, Part B*, **51B**, 1925 (1996).

201. L. C. Azeredo, R. E. Sturgeon, and A. J. Curtius, *Spectrochim. Acta, Part B*, **48B**, 91 (1993).

202. E. Beinrohr, M. Cakrt, M. Rapta, and P. Tarapci, *Fresenius J. Anal. Chem.*, **335**, 1005 (1989).

203. M. Wang, A. I. Yuzefovsky, and R. G. Michel, *Microchem. J.*, **48**, 326 (1993).

204. D. B. Do Nascimento and G. Schwedt, *Anal. Chim. Acta*, **283**, 909 (1993).

205. R. Ma, W. Van Mol, and F. Adams, *Anal. Chim. Acta*, **293**, 251 (1994).

206. Y. An, S. N. Willie, and R. E. Sturgeon, *Spectrochim. Acta, Part B*, **47B**, 1403 (1992).

207. I. L. Shuttler, M. Feuerstein, and G. Schlemmer, *J. Anal. Atom. Spectrom.*, **7**, 1299 (1992).

208. H.-W. Sinemus, H.-H. Stabel, B. Radziuk, and J. Kleiner, *Spectrochim. Acta, Part B*, **49B**, 643 (1993).

209. H.-W. Sinemus, J. Kleiner, H. H. Stabel, and B. Radzuik, *J. Anal. Atom. Spectrom.*, **7**, 433 (1992).

210. O. Nygren, C.-A. Nilsson, and W. Frech, *Anal. Chem.*, **60**, 2204 (1988).

211. O. Røyset and Y. Thomassen, *Anal. Chim. Acta*, **188**, 247 (1986).

212. K. Wróbel, E. B. Gonzalez, K. Wróbel, and A. Sanz-Médel, *Analyst*, **120**, 809 (1995).

213. *Instructions for the Use of Digestion Vessels*, CEM Corporation, Indian Trial, NC, 1990.

CHAPTER 7

PRACTICAL HINTS ON THE DETERMINATION OF ELEMENTS BY GFAAS

The goal of this chapter is to provide some guidelines for quantitative analysis by GFAAS. First, the criteria that are used to evaluate whether GFAAS can be used to do a particular analysis are outlined. The second section discusses sampling, storage of samples, and sample preparation. The emphasis is on contamination, which is a common source of error in trace analysis. The use of quality control procedures is discussed to evaluate analytical procedures, including the use of standard reference materials and recovery checks. The instrument optimization protocols required to do quantitative analysis by GFAAS are discussed, such as pyrolysis temperature optimization, atomization temperature optimization, and the type and quantity of chemical modifier. The final section of this chapter is a troubleshooting guide for GFAAS analysis.

7.1 APPLICABILITY

The applicability of an elemental analysis technique involves consideration of the analyte, the amount of available sample, and the concentration levels of the analyte. The first criterion involves consideration of the applicability of GFAAS to determine a particular element. In general, GFAAS is applicable to the determination of most metals and metalloids, with the exception of a few refractory elements (e.g., tantalum). The atomic absorption cookbook provides a list of determinable elements.

The amount of sample must also be considered. An advantage of GFAAS compared to other techniques is the small amount of sample required. Each graphite furnace cycle employs approximately 10 to 20 μL of solution (dissolved sample or liquid), and since determinations are normally performed in triplicate, approximately 30 to 60 μL are required. If a smaller volume is available, but the analyte levels are relatively high, it is possible to dilute the sample before analysis to increase the volume. Slurry and solid sampling methods may employ as little as 1 to 4 mg of solid per cycle, allowing analysis of a few milligrams of sample, although such small masses may not be representative of the bulk of the material.

Assuming the analyte is determinable by GFAAS, and sufficient sample is available, the third criterion to consider is the concentration levels in the sample following sample preparation procedures (Chapter 6). Generally 0.1 to 1 g of solid sample are dissolved and diluted per 100 mL volume. The useful (linear) range of the calibration graph is usually assumed to be between the limit of quantitation (approximately five times the detection limit) and the level of linearity (Section 3.2), typically two to three orders of magnitude. It may be possible to detect values closer to the detection limit, but degraded precision and accuracy should be expected. In addition, some sample matrices may degrade the detection limit, increasing the limit of quantitation. Data from the atomic absorption cookbook and the approximate concentration of analyte in the sample should be used to determine whether the levels fall within the useful range of the graph.

Obviously, if this condition is met, the analyst may continue to the next step. However, if the concentration levels are too low, the analyst has two options. The easiest option, if available, is the use of a more sensitive technique. This may not be possible because GFAAS is one of the most sensitive techniques. Possible options include inductively coupled plasma–mass spectrometry and neutron activation analysis. The second option is to use one of the preconcentration techniques (e.g., extraction, chromatography, flow injection) described in Section 6.4. These techniques also offer the advantage of separating the analyte from the matrix, which may reduce interferences. Their primary disadvantages are their time-consuming nature and inconvenience.

If the concentration levels are above the useful range of the calibration curve, several options are available. First, if the concentration levels are sufficiently high, the use of a less sensitive technique, such as flame AAS or inductively coupled plasma optical emission spectrometry, is appropriate. These techniques are faster and usually easier than GFAAS. A second option is dilution of the sample by deionized-distilled water, which has the advantage of diluting potential interferences. Third, many elements have less sensitive alternative wavelengths, listed in the cookbooks, that may be employed to determine relatively high concentration levels. A final option is the use of a low internal flow of gas through the graphite tube during the atomization step, which serves to more quickly remove the atoms from the atomizer and reduce the sensitivity. This option should probably be used as a last resort because a gas flow may reduce the temperature during the atomization step and cause chemical interferences.

7.2 SAMPLING, SAMPLE STORAGE, AND SAMPLE PREPARATION

The use of well-designed sample collection and storage procedures is required to ensure collection of representative samples with good precision and accuracy. A number of recent volumes and chapters have discussed sampling for trace element analysis (1–5). The reader is referred to these references for sampling and storage procedures for specific types of samples, for example, clinical (1,2), seawater (6,7), plant samples (5), and environmental samples (4).

In order to obtain a representative sample, it is first necessary to consider the size of the gross sample required that is truly representative of the entire sample. It is then necessary to reduce the gross sample to laboratory samples, which are employed for analysis, with maintenance of the chemical integrity of the analytes. Ideally, in addition to ensuring that the laboratory sample is representative of the entire sample, it is necessary to ensure that no addition or deletion of analyte has occurred prior to analysis. Reduction of the sample size involves homogenization of the sample by thorough mixing. Some samples, such as soils and fertilizer blends, are heterogeneous. In these cases the particle size should be reduced as much as possible in order to obtain representative portions for analysis. Particle size reduction of hard materials may be performed with laboratory mills and grinders. Soft samples, such as foods and tissue, may be homogenized with mixers or blenders.

It should be pointed out that in some cases it may not be desirable to homogenize the entire sample. For example, consider fruits with nonedible skins (8). The levels of metals may be higher in the skins than in the fruit due to pollution. However, is the concentration in the skin of interest? It may be more appropriate to analyze the edible portion.

Loss of analyte may occur during transport, storage, and sample preparation. For example, the analyte may coprecipitate with other salts (e.g., urine samples) (2). These losses may be eliminated by complete digestion of the sample. The analyte may absorb on to the wall of a container (3). Absorption losses can be minimized by the use of thoroughly cleaned Teflon or polyethylene containers, acidification of the sample to $pH < 1$, and minimization of the contact time. Although volatilization of elements is normally associated with sample preparation procedures (Chapter 6), losses of volatile compounds (e.g., mercury compounds) may occur at room temperature.

Contamination is a particularly significant factor in GFAAS because of the relatively low (pg/mL to ng/mL) concentration levels determined. Some sources of contamination for individual elements are given in Table 7.1. Air particulate matter (dust) may be a major source of contamination, both at the sampling location and in the laboratory. When collecting plant samples, it is often desirable to monitor the concentration of metals in tissues independent of the air particulate matter (5). In this case it is necessary to wash the samples to remove the deposition, which may lead to additional contamination/losses.

The prevention of contamination by air particulate matter may be achieved by the use of clean environments. Class 100 clean rooms appear to be adequate at removing most particulates. Ideally, the air entering the clean room should be

Table 7.1 Sources of Contamination for Elements (8)

Element	Sources of Contamination
Aluminum	Dust, glass, low-pressure polyethylene, PVC, laboratory tissue paper, agate (mortar and pestle), aluminum ceramic (for grinding samples)
Antimony	Glass
Boron	Boron carbide (for grinding samples)
Beryllium	Laboratory tissue paper
Cadmium	Plastic
Calcium	Polyethylene, polypropylene, PVC, dust, agate (mortar and pestle), aluminum ceramic (for grinding samples)
Chromium	Stainless steel
Cobalt	Stainless steel, anticoagulants for blood, tungsten carbide (for grinding samples)
Copper	Polyethylene, plastic, anticoagulants for blood, steel
Iron	Polyethylene, steel, dust, agate (mortar and pestle)
Lead	Exhaust fumes, glassware
Manganese	Stainless steel
Magnesium	Glassware, agate (mortar and pestle), aluminum ceramic (for grinding samples), dust
Nickel	Stainless steel
Potassium	Teflon, agate (mortar and pestle)
Silicon	Dust, agate (mortar and pestle), fused zirconia (for grinding samples), aluminum ceramic (for grinding samples)
Sodium	Analyst's skin, dust, polyethylene, polypropylene, PVC, Teflon, agate (mortar and pestle)
Strontium	Glassware
Tungsten	Tungsten carbide (for grinding samples)
Titanium	Polyethylene, tungsten carbide (for grinding samples)
Zinc	Dust, rubber, anticoagulants for blood, polyethylene, polypropylene
Zirconium	Fused zirconia (for grinding samples)

purified with high-efficiency particulate air (HEPA) filters, which serve to remove at least 99.99% of 0.3-μm particles. Woittiez and Sloof (3) described actions required to convert an ordinary laboratory into a clean environment, which include the removal of metal objects, painting surfaces with epoxy paint, and the placement of sticky mats near the entrance.

The analyst may also be a source of contamination. Human skin is a source of sodium and other elements. Significant contamination may also be produced by hair and clothing. Woittiez and Sloof (3) recommend limiting access to the clean room and requiring special dust-free clothes, shoes, and hats.

Equipment and chemicals are potential sources of contamination. Collection and storage should be performed using clean containers (washed in detergent followed by soaking in 1*M* nitric acid) made of a high-purity material in a relatively clean area with controlled temperature conditions appropriate for the samples. For trace element analysis, recommended materials include polyethylene, Teflon, and synthetic quartz. Sample vessels should be permanently labeled to allow random assignment to prevent bias from particular containers. Colorless pipette tips should be used for solution preparation because the color of some tips is due to the presence of certain metals.

Homogenization of samples by grinding or blending induces considerable physical contact between the sample and this equipment. Considerable contamination may be induced during this step in the sample preparation procedure (Table 7.1). The choice of material to be used for grinding is dependent on the analytes. For example, steel is a durable, relatively inexpensive material but may induce contamination of iron, chromium, and manganese, while tungsten carbide is brittle and expensive but elemental contamination is limited to tungsten, cobalt, and a few rare-earth elements.

High-quality deionized water is essential for trace element analysis. A number of water purification systems are available commercially. Further purification of water using a sub-boiling distillation unit may be necessary for extremely low concentration levels.

The use of high-purity chemical reagents is obvious for trace analysis. Contamination may be a problem even with the use of high-quality reagents, as illustrated in Table 7.2 (1). The levels of copper and zinc in blood samples are typically in the microgram per milliliter range, and contamination from acids is not a problem. However, the other elements are present at the picogram per milliliter to nanogram per milliliter levels, and concentrations in the reagents are the same as or greater than those in the samples. These data indicate that it may be necessary for the laboratory to purify acids for trace analysis. Quartz sub-boiling distillation units may be used to reduce metal concentrations in all acids except hydrofluoric, which must be purified in Teflon units. Details on sample preparation procedures are given in Chapter 6.

The significance of contamination control is illustrated in Table 7.3 (1), where the determination of manganese in serum by neutron activation analysis is reported in the absence and presence of contamination controls. Contamination controls for this work involved the use of care in sample handling, carefully cleaned glassware and plasticware for the collection and storage of samples, purified reagents, and a clean room. The concentration levels reported here are of the same order of magnitude as would be expected by GFAAS, and similar results would be expected with this technique. The manganese levels are a factor of 10 lower in the absence of contamination, and the RSD was reduced from approximately 100 to 15%. These data illustrate the concept of "garbage in, garbage out" for elemental analysis. Reliable data cannot be obtained during the analysis step if errors are introduced in the collection and preparation steps.

Table 7.2 Estimated Concentration Levels in Human Blood Plasma or Serum of Several Biological Elements and Maximum Levels of Impurities[a]

Element	Plasma or Serum Level, μg/mL	Acids, Maximal Levels of Impurities, μg/mL		
		Nitric	Perchloric	Sulfuric
Ag	0.00020	≤0.0014	≤0.0033	≤0.0018
Al	0.0035	≤0.007	≤0.0084	≤0.0092
As	0.0010	≤0.0014	≤0.017	≤0.0092
Cd	≤0.00025	≤0.0014	≤0.0017	≤0.0018
Co	0.00020	≤0.0014	≤0.0017	≤0.0018
Cr	0.00015	≤0.0028	Not reported	≤0.0018
Cu	1.0	≤0.0014	≤0.0017	≤0.0018
Mn	0.00055	≤0.0014	≤0.0017	≤0.0018
Mo	0.00055	≤0.0014	≤0.0017	≤0.0018
Ni	0.00025	≤0.0070	≤0.0033	≤0.0037
Sn	0.00050	≤0.0070	≤0.0084	≤0.0018
V	0.000040	≤0.0014	≤0.0017	≤0.0018
Zn	0.90	≤0.0070	≤0.0084	≤0.0092

[a] As specified in 1990 by the manufacturer (Merck) in Suprapur® Nitric, Perchloric, and Sulfuric Acids (1)

Table 7.3 Manganese Concentrations (ng/mL) Determined by Neutron Activation Analysis in Duplicate Serum Samples from 12 Volunteers with and without Contamination Control Procedures (1)

Individual Monitored	Without Contamination Controls		With Contamination Controls	
	Sample (1)	Sample (2)	Sample (1)	Sample (2)
1	1.3	1.8	0.56	0.58
2	4.6	2.3	0.64	0.44
3	2.3	2.8	0.64	0.57
4	12	29	0.58	0.51
5	3.8	3.4	0.60	0.60
6	1.9	8.5	0.72	0.60
7	5.7	3.1	1.02	1.06
8	2.8	5.7	0.44	0.49
9	24	6.3	0.57	0.70
10	2.1	9.4	0.65	0.81
11	12	6.1	0.54	0.55
12	7.8	4.4	0.64	0.78
Mean ± SD	6.7 ± 6.6	6.9 ± 7.4	0.63 ± 0.10	0.64 ± 0.14

This trend has been confirmed in several analyses in the environmental (e.g., seawater) and clinical chemistry (e.g., blood serum) literature in which ambient levels of metals have "fallen" over the past 30 years because of the elimination of contamination in more recent work.

Benoit et al. (9) determined cadmium, copper, lead, and silver in freshwaters by GFAAS. Clean techniques were required for reliable determination of cadmium, copper, and lead. Contamination was not a problem for silver, which was classified as a "rare" element. The authors recommended the use of clean techniques, even for the measurement of metals in polluted waters, which have moderately high concentration levels.

Woittiez and Sloof (3) suggest a presampling plan must be assembled prior to execution of any part of an analytical procedure (Table 7.4). This may prevent the analyst from undertaking sampling, sample preparation, and analysis that does not answer the research question and hence is a waste of time. After the

Table 7.4 Presampling Plan

Question	Comments
1. Has the research question been adequately formulated?	If no, sampling, sample preparation, and analysis are a waste of time.
2. Has the specific analysis been defined?	If no, sampling, sample preparation, and analysis are a waste of time.
3. Can the analysis be performed using state-of-the-art analytical procedures and instruments?	If no, sampling, sample preparation, and analysis are a waste of time, unless research is performed to develop the required methods.
4. Does the laboratory have the expertise and equipment to perform the specific analysis?	If no, assistance from an expert laboratory is required or sampling, sample preparation, and analysis are a waste of time.
5. Has the sample been defined and is it possible to collect this sample?	If no, sampling, sample preparation, and analysis are a waste of time.
6. To solve the research question, is it necessary to consider the sample be representative for a bulk sample or population?	If yes, it is necessary to obtain information about the homogeneity of the analyte in the sample and uniformity of the analyte in the population.
7. Can a representative sample be collected?	If no, final results are restricted to the sample collected.
8. Can the integrity of the sample be guaranteed during sampling and transport?	If no, analysis is a waste of time.

Adapted from Woittiez and Sloof (3).

presampling plan has been evaluated, a sampling and sample preparation plan should be developed describing all steps performed prior to analysis (Table 7.5).

Table 7.5 Sampling Plan

1. Sampling	
1.1. What is sampled?	1.2. When is sampling performed?
1.3. Where is sampling performed?	1.4. How is sampling performed?
1.5. Is special equipment required?	1.6. Who collects the samples?
1.7. How many samples are collected?	1.8. What size samples are collected?
1.9. How are the samples coded?	1.10. Is a composite sample required?
1.11. Are samples directly contained for transport?	1.12. What containers are employed?
1.13. Can contamination and losses be avoided?	1.14. Is in situ analysis necessary?
1.15. Is in situ cleaning necessary?	1.16. Are geographic and meteorological data required?
1.17. Which data are recorded?	1.18. How are the data recorded?
2. Subsampling and Packing	
2.1. Is in situ subsampling required?	2.2. Is in situ repacking required?
2.3. Which containers are used?	2.4. Are the containers clean?
2.5. Can contamination and losses be avoided?	2.6. Are containers coded?
2.7. Is in situ physical preservation required?	2.8. Can contamination and losses be avoided in the preservation step?
2.9. Is in situ physical preservation required?	2.10. Is special equipment required?
2.11. Which data are recorded?	2.12. How are the data recorded?
3. Transport and Sample Preparation	
3.1. Is there a deadline for sample transport?	3.2. Is storage space available?
3.3. Is immediate preparation required?	3.4. Are special conditions in lab required?
3.5. Is sample cleaning required?	3.6. Is sample division needed?
3.7. Is sample reduction required?	3.8. Is drying required?
3.9. Is crushing required?	3.10. Is milling required?
3.11. Is sieving required?	3.12. Is mixing required?
3.13. Can contamination and losses be avoided in all these steps?	3.14. Is immediate analysis required?
3.15. Is the laboratory sample stable?	3.16. Which data are recorded and how?

Adapted from Woittiez and Sloof (3).

7.3 QUALITY CONTROL PROCEDURES

Quality in analytical procedures is characterized by the magnitude of errors and the extent to which the errors affect the final results (8,10,11). Accredited laboratories are required to document the accuracy and precision of methods and results as described by international organizations such as the International Standards Organization (ISO), International Union of Pure and Applied Chemistry (IUPAC), and Association of Official Analytical Chemists (AOAC). All samples and procedures must be carefully documented during collection, storage, and analysis.

Careful attention must be paid to blanks and calibration standards. Calibration and reagent blanks should be prepared and analyzed to establish a zero baseline and a background value, respectively. Generally two independent sets of high-quality calibration standards should be employed. Calibration standards are generally prepared by serial dilution of concentrated stock solutions (1000 μg/mL). Directions for preparation of stock solutions are provided with the list of standard methods provided with the instrument in Table C.3, or, alternatively, commercial standards may be purchased. Stock and calibration standards are usually stored in acid solution in plasticware to increase their stability. Analytical standards for GFAAS should be prepared daily in plasticware (rather than glassware) in dilute nitric acid (0.2%) by serial dilution techniques. Generally dilutions should be performed with mechanical pipettes with volumes between 0.1 and 5 mL.

Standard reference materials (SRMs), which are samples that have been analyzed by at least two independent methods, should be analyzed with the "real" samples to assess accuracy (12–14). Watters (14) proposed three criteria for the selection of SRMs for a particular analysis. First, the SRM should have a matrix composition similar to the sample. Second, the analyte should be at a concentration level in the SRM similar to the expected level in the sample. Third, the uncertainty of the SRM certified concentration should be lower than the specified level of bias for the analysis.

A wide variety of standard reference materials are available, such as manufactured matrix (e.g., plant process quality control for metals, cement, glass, paint, and automobile catalysis), agricultural products, environmental (soil, sediment, sludge, and water), botanical, marine science, geological, coal, and medicine (12,13). Some government agencies that supply SRMs are listed in Table 7.6. A number of private companies also produce reference materials that are listed in the annual (August 15) laboratory guide of *Analytical Chemistry*.

Recovery checks involve the addition of an aliquot of analyte to a real sample to evaluate the recovery efficiency of the method. Recovery checks should be incorporated randomly in a sequence of analyzed samples. The level of spiked analyte may be equal to the expected level of analyte in order to evaluate

Table 7.6 Representative Government Suppliers of Standard Reference Materials (12,13)

Agency	Field of Science
Community Bureau of Reference (BCR), Brussels, Belgium	Numerous
National Institute of Standards and Technology (NIST), Gaithersburg, MD	Numerous
National Research Council (NRC), Montreal Road, Ottawa, Ontario, Canada	Marine
National Institute For Environmental Studies (NIES), Ibaraki, Japan	Environmental
Geological Survey of Japan (GSJ), Ibaraki, Japan	Geological
United States Geological Survey (USGS), Denver, CO	Geological
Laboratory of the Government Chemist (LGC), Middlesex, UK	Numerous
Canadian Certified Reference Materials Program (CCRMP), Ottawa, Ontario, Canada	Geological
Agricultural Research Center (ARC), Jokioinen, Finland	Agricultural
National Research Center for Certified Reference Materials (NRCCRM), Beijing, China	Numerous

differences between analyte from the sample or from the spike. On the other hand, minimal error in the recovery factor is obtained when the added analyte is several times larger than the native analyte. It is necessary to ensure that the total concentration of analyte is on the linear portion of the calibration graph. The use of recovery checks in quantitative trace analysis is the subject of a volume edited by Parkany (15).

Samples should be analyzed using strict quality control procedures. *Blind samples* are standards that are submitted for analysis as real samples. Analysis of replicates involves repeated analysis of a sample during a series of measurements in order to evaluate the precision of the analytical system. Typically at least one sample should be replicated in every ten analyses.

Shewhart control charts, which involve a plot of the mean measured concentration of control samples versus time, are used to monitor long-term precision and accuracy of an analytical method (Fig. 7.1). They typically consist of a center line, which represents the target value, with symmetrically arranged limits defined in terms of σ_p, where $\sigma_p = \sigma_Y/\sqrt{n}$, with σ_Y is the standard deviation of the analysis and n is the number of replicates. Generally $\pm 2\sigma_p$ serves as a warning limit and $\pm 3\sigma_p$ as an action limit.

In addition to quality control within a laboratory, it is also necessary to verify comparability between laboratories. External quality assurance systems are employed to assess the reliability of results from more than one laboratory.

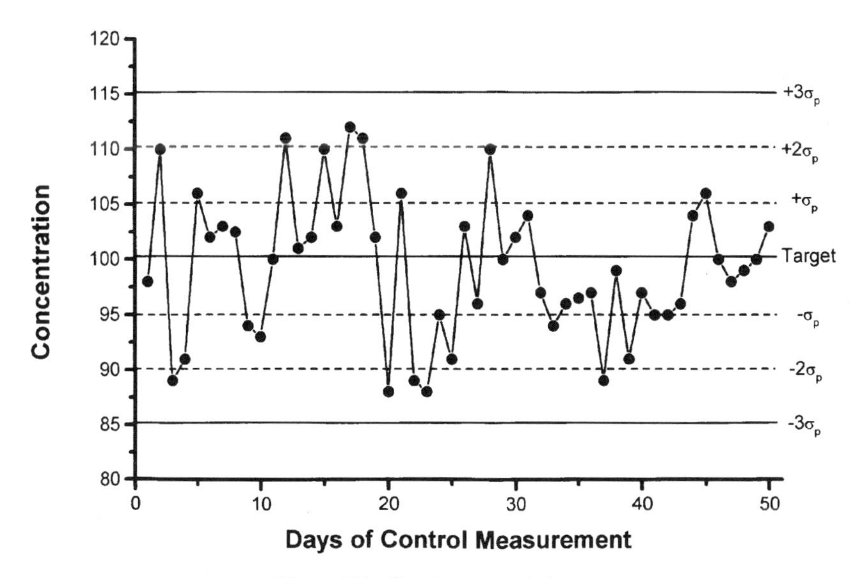

Figure 7.1. Shewhart control chart.

Further details on quality control procedures are available in References 10 and 11.

7.4 DEVELOPMENT OF GFAAS METHODS

The successful use of GFAAS for real sample analysis involves the use of modern furnace technology (Table 1.1), as we have emphasized throughout this volume. Method development should be initiated by consultation with an atomic absorption cookbook of experimental conditions for the determination of a particular element, provided by most manufacturers. Methods are usually outlined for the determination of elements in particular samples, which may include the concentration levels that correspond to the linear region of the calibration graph, sample preparation procedures, instrument conditions, such as available wavelengths, with their relative sensitivities, bandpasses, and temperature programs, and type and amount of chemical modifiers. A typical example of this information is provided in Table 7.7. However, it is the opinion of the authors that the analyst should use these conditions as general guidelines and develop their own sample preparation methods and instrument conditions. Additional information may be obtained by careful examination of the atomic spectrometry literature before attempting an analysis to avoid "reinventing the

Table 7.7 Typical Atomic Absorption Cookbook Information (28)

Application: Sb in Forensic Samples

Element	Sb		
Mode:	Single Beam Absorption		
Integ. Type:	Peak Area	Integ Time: 6.0 s	Delay Time: 0.0 s
Graph Scale:	Automatic		
Comment:	Platform atomization, 2 μL of 100 ppm Pd + 200 ppm Mg as $Mg(NO_3)_2$ as modifier		

Furnace Program

	Dry Step	Pyrolysis 1 Step	Pyrolysis 2 Step	Atom Step	Clean Step
Temp, °C	200	650	750	2300	2350
Ramp, s	2	15	10	0[a]	1
Hold, s	3	10	5	4	2
Purge gas flow	Low	Medium	Medium	Off	Medium

Integ. Stage:	Atom	Air Ash: Off		Auto-Baseline Corr: Yes
Wavelength:	217.60 nm			
Bandpass:	0.4 nm			
Background Method:	Smith-Hieftje			
Lamp Current:	Normal:	6.0 mA		Background: 2.5 mA
Significant Figures:	4		Print Units:	ppb
Stock Standard Conc.	50			

Standardization Method: Aqueous Calibration

Standard Name	Standard Concentration, ppb
BLANKSTD	0
10STD	10
20STD	20
30STD	30
40STD	40
50STD	50

Standardization Method: Standard Additions

Standard Addition Name	Standard Addition, ppb
15STDADD	15
25STDADD	25
50STDADD	50

[a] Maximum power heating.

wheel." Methods of sample preparation are discussed in Chapter 6 and additional applications are listed in Appendix B. It is recommended that the reader focus on references published since the mid-1980s, subsequent to the development of modern furnace technology, for the most relevant information.

A number of GFAAS instrumental conditions need to be selected or optimized. For example, a method of background correction should be selected if more than one is available on an instrument (Section 4.6). Many modern instruments have self-reversal or Zeeman and the continuum source method. In general, self-reversal or Zeeman is preferable to the continuum source method, although there are exceptions to this statement (Section 4.6.5).

The type of graphite must be selected as well. A pyrolytically coated graphite tube (Section 4.2.1) with platform atomization (Section 4.2.3) gives optimum performance for most elements. However, if cost of graphite is an important consideration, some volatile elements may be accurately determined with less expensive uncoated graphite tubes, and some involatile elements may be accurately determined without a platform. The cookbook should be consulted for the type of graphite recommended for a particular application.

The use of chemical modifiers (Section 5.1.2) has been shown to reduce chemical interferences for GFAAS. The atomic absorption cookbook and literature should be investigated for the most appropriate choice of chemical modifier for a given analyte. For example, palladium has been commonly used to determine a variety of volatile elements because of its ability to stabilize them sufficiently to allow pyrolysis temperatures above 1000°C and remove much of the matrix. Frequently several modifiers have been employed for a given element, and it is desirable to experimentally evaluate the most suitable. After the modifier has been selected, it is necessary to optimize the amount of reagent employed. Chemical interferences caused by organic matrices have been effectively removed by the use of oxygen ashing at temperatures below 800°C to prevent oxidation of the tube.

Lastly, the atomization cycle of the graphite tube needs to be optimized for the particular analysis (Section 4.2.2). Most cookbooks provide temperature programs that can be used as a starting point. In general, conditions for the dry cycle are determined by the graphite employed (e.g., whether a platform is present or not), and hence they usually do not need to be optimized for each analysis. The pyrolysis and atomization temperatures are optimized for aqueous standards and the samples (Fig. 4.10). Ideally, no difference would be observed in the optimum temperatures for standards and samples. The characteristic mass or limit of detection should be determined and compared to values specified in the cookbook to evaluate the analytical performance of the system.

Quality control procedures should then be employed to evaluate the precision and accuracy of the analysis (Sections 7.2 and 7.3). These procedures include the sample collection methods, including contamination control, the use of high-quality standards, and the use of standard reference materials and recovery checks.

From the discussion above, it should be apparent that a number of experiments must be performed in order to optimize conditions and do an analysis by GFAAS. Normally one might expect to spend several days performing optimizations for the determination of an element in a "new" sample.

7.5 TROUBLESHOOTING

As mentioned earlier, it is necessary to develop sampling, sample preparation, and GFAAS analysis protocols for each analyte and sample. Manufacturer's books and recommended values are good starting points for GFAAS methods, and the values and suggestions are usually close to the optimum. This section presents some practical hints and/or suggestions in developing experimental and instrumental conditions as well as some practical troubleshooting hints. The manufacturer's operators manual will most likely have a section on troubleshooting the instrument.

The Furnace Will Not Heat. Modern graphite furnaces are equipped with a number of safety devices to prevent the graphite furnace from overheating. First, it is essential that cooling water flow through the furnace at the recommended flow rate. This may be achieved by tap water from a faucet or by a water recirculator. The latter system is probably preferred because deionized water may be placed in the water recirculator, which minimizes buildup of deposits in the furnace head. Occasionally mold will accumulate in the cooling system; pumping 25% Clorox through the system will usually eliminate this problem. After deposits have formed, it is a good idea to momentarily reverse the direction of flow to initiate their removal. Second, the purge gas (usually argon) must be above a minimum pressure level. It is a good idea to check the pressure of argon tanks at the beginning of each day of analysis to avoid running out of argon while the system is operating. Third, the furnace has a temperature sensor that shuts off current from the power supply if the temperature exceeds a maximum value. If the flow of water through the cell head, or long furnace cycles with high temperatures are employed, the temperature sensor may shut off heating to the furnace. Fourth, the graphite tube will not heat unless good physical contact is present between the tube, the graphite electrodes, and the water-cooled brass electrodes. For example, if the tube cracks, then the furnace typically will not heat. The authors recommend weekly cleaning of the inner surface of the graphite contact into which the graphite tube is inserted using 70% ethanol/water with a cotton swab.

Intensity of the Source Is Diminished, Erratic, or Nonexistent. Generally this means that the hollow cathode lamp has failed. The first thing to try is to turn off the power to the lamp, remove the lamp, and insert a new one. Another thing to check is the cleanliness of the furnace windows. They should be examined regularly (daily if large quantities of samples are analyzed) and cleaned with 70% ethanol/water with lens paper.

Temperature Obtained During Maximum Power Heating Is Higher Than the Setting. An optical pyrometer is used to monitor the temperature of the tube during maximum power heating of the tube. In some instruments, the sensor is below the tube and is protected from degradation by matrix components with a quartz window. A dirty window will cause the furnace to heat to higher temperatures than entered by the operator. Other instruments require manual calibration of the temperature obtained during maximum power heating. This calibration should be performed when the atomization temperature is varied.

When Should the Graphite Tube Be Replaced? Normally, it is a good idea to monitor the number of furnace firings. Typically one can expect 200 to 500 atomization cycles. Higher numbers are expected with low atomization temperatures, low concentrations of reagents (e.g., acids), and dilute matrices. The tube should be visually inspected at the beginning of operation each day. If the shiny coating on the tube has worn off, it is probably necessary to replace the tube. With a pyrolytically coated tube, the removal of the pyrolytic coating is usually easy to observe. Alternatively, the tubes may be used until degradation in the precision (>5–10% RSD) is observed.

Graphite Tubes Seem to Wear Out Faster Than Expected. As discussed, normally a graphite tube may be used for 200 to 500 atomization cycles, depending on the temperatures and matrix employed. Some modifiers and acids, such as lanthanum and sulfuric acid, induce rapid degradation of tubes. The presence of oxygen causes rapid tube wear at temperatures above 700°C. Some furnace systems enclose the furnace in a chamber with a door that opens and closes for sample introduction. The door must close properly to prevent oxygen from entering the chamber. Oxygen ashing (Section 5.1.2) is an effective method to remove organic matter from the tube, but temperatures below 700°C should be employed to prevent rapid deterioration of tubes. A tank of argon with high amounts of oxygen may induce rapid degradation of the furnace.

A potential problem in GFAAS is the quality of the graphite surface and graphite tube. Many laboratories report conflicting results: Some claim they can use a high tolerance of strong acids while others report a low tolerance. It appears that the quality of the graphite tubes (surfaces) is variable, from batch to batch, and even within batches, resulting in different rates of wear. Commercial manufacturers have recognized this problem and have taken steps to ensure the availability and quality control of the graphite tubes. However, variation in analysis conditions between different laboratories with the same instrument continue to be commonplace. For this reason, it is essential that each laboratory optimize its GFAAS analysis conditions.

A poor graphite surface can give memory effects, multiple peaks, poor precision, and reduced sensitivity. This can be caused by heavy use of the graphite cuvette, use of high temperature for extended time intervals [determination of refractory elements such as molybdenum require the high temperatures (2700°C) for atomization times up to 10 s], use of strong oxidizing agents such as 5% nitric acid (frequently used in digestion procedures), leaks in the atomization cell, poor electrical contact between the tube and graphite electrodes, or poor quality coating. It may be difficult and time consuming to pinpoint the problem and is best solved by replacing the graphite tube with a new, high-quality graphite tube.

There Is No Signal. This may be caused by a number of problems. A bad HCL may appear to work but not emit radiation that is absorbable by atoms. An HCL known to be operational should be tested to investigate if this is the cause of the problem. The graphite tube should be examined visually for damage. If the tube seems worn, it should be replaced. The use of a pyrolysis temperature that is too high will vaporize the analyte before the atomization step, resulting in no signal. Similarly, the use of an atomization temperature that is too low (higher temperatures are required with some chemical modifiers) may not atomize the analyte. The operation of the autosampler should be investigated. Using a dentist's mirror or an in-board video camera, watch the autosampler deliver into the tube to make sure it is delivering the sample to the platform or tube.

Usually it is a good idea to have a "favorite" element to confirm that the overall operation of the instrument is satisfactory. An easy element, such as copper, should be selected to confirm that the instrument is operational when problems arise with an unfamiliar analyte or sample.

Signal Is Small or the Precision Between Furnace Firings Is Poor. The continuous determination of a solution should give a precision of between 1 and 5% (depending on the concentration of the element and the sample matrix). Many of the problems that may cause no signal may also induce a small signal or poor precision: failure of the HCL, a worn tube causing absorption of analyte, excessively high pyrolysis temperature, excessively low atomization temperature, misalignment of the autosampler. Poor precision between atomization cycles may also be caused by improper drying of the sample. A high dry temperature at fast heating rate may induce splattering of the sample that may lead to irreproducible loss of sample. It is also necessary to make sure that the internal flow of purge gas is turned off during the atomization step to prevent rapid removal of analyte from the tube.

How Do I Check the Alignment of the System? Normally the first step is to align the graphite tube with the slit of the monochromator. Darken the room slightly and open the cover on the detection system so that you may observe the slit. Manually heat the tube to approximately 2000°C so that the image of the furnace is around the slit. You should also move the furnace or a focusing device (lens) in the axis of the optical path to sharpen the image. This alignment is important to minimize the quantity of blackbody emission that reaches the detection system. High levels of emission may degrade precision and saturate the photomultiplier tube (PMT) at high atomization temperatures. Now adjust the position of the HCL so that the maximum amount of light is transmitted through the tube. The location of the adjustment devices is dependent on the manufacturer of the instrument, and the manual should be consulted for details.

How Do I Set Up the Autosampler? The autosampler should be mounted on to the instrument according to the instrument manual. The autosampler arm should be adjusted so that it passes through the center of the dosing hole and stops $\sim\frac{1}{16}$ inch above the tube or platform. We recommend setting it manually each day, and then carefully observing the motion and delivery for the first two to three furnace cycles.

When Should the Method of Standard Addition Be Employed? As discussed in Section 3.3, standard addition is only effective at correcting multiplicative interferences, such as chemical interferences (Section 5.2). Standard addition will not correct for additive interferences, for example, background (16).

How Do I Know Whether My Determinations Are Accurate? The first step is to use good quality control procedures (e.g., sampling, sample preparation, use of SRMs, recovery checks, etc.). The biggest question is in the accuracy of the background correction procedure. First, the background level should be minimized as much as possible by use of chemical modifiers and pyrolysis steps. In general, it is desirable to use as high a pyrolysis temperature as possible to remove as much of the matrix before the analytical measurement. In extreme cases, it may be necessary to isolate the analyte from the matrix using a preconcentration/isolation procedure (e.g., extraction, chromatography, flow injection; Section 6.4). Second, if available, either Zeeman or self-reversal background correction are preferable to continuum source correction. The first two methods correct at or very near the analytical wavelength, reducing the number of errors, and generally can accurately correct for higher background levels. Evaluation of the effectiveness of the background correction may be difficult. Overcorrection errors are easy to observe due to the negative corrected signal obtained (Fig. 4.22). Undercorrection is much more difficult to spot. If the analyte concentration is sufficiently high, analysis at various dilution levels should give identical values if the background correction system is operating correctly. Alternatively, if a second analytical wavelength is available (Appendix C), the analysis may be performed at that wavelength. If a neighboring line that is not absorbed by the analyte is available, do the measurement at that wavelength with the background correction. If the background correction is working correctly, no signal should be present. If available, it may be advisable to do the measurements with more than one background correction system and compare the results. Finally, it is advisable to consult reported interferences in Tables 4.3 and 4.4. For example, some chemical modifiers have caused interferences for specific elements.

How Is the Display of Absorption and Background Signals Used? These are important diagnostic tools for analysis. First, you should make sure that the corrected signals of standards and samples are present within the selected integration time. In addition, you should minimize the size of the window to just include all of the signals. Integration for longer periods than necessary degrades the signal-to-noise ratio and precision (17). Second, ideally the signal should be a single, relatively narrow peak. The presence of double peaks may indicate atomization from more than one surface (e.g., if the autosampler is delivering sample to both the platform and wall).

How Is the Zero Absorbance Measurement Made? Accurate measurement of zero absorbance is required to obtain good accuracy. Some instruments use data points collected just prior to atomization to obtain an average value that is subtracted from all points in the integrated profile. This procedure, which is called *baseline offset compensation* (BOC), has the advantage of compensating for drift between individual atomization cycles. Barnett et al. (17) recommend the use of BOC times of 2 to 4 s to obtain optimum precision. Alternatively, an "autozero" measurement is made in which absorbance is recorded without a furnace firing. This method does not allow compensation with every furnace measurement, and hence drift may occur between autozero measurements.

How Are Memory Effects Eliminated? Memory effects involve retention of analyte from one furnace firing to another. For relatively volatile elements, memory effects can be alleviated by use of a cleaning step subsequent to atomization. Homogeneous refractory elements (e.g., tungsten, boron) may remain in the graphite tube even after a cleaning step. It may be necessary to run dry atomization cycles between each standard or sample measurement and observe a low (ideally zero) integrated absorbance value to ensure that memory effects are not present. Memory effects may be caused by high levels of analyte present on the graphite electrodes. In the authors' experience, it is generally necessary to replace the graphite electrodes to eliminate this problem.

Why Are Multiple Peaks Present? One source of multiple peaks is memory effects (see above). Multiple peaks can be caused by several factors including too low a drying temperature or too short a drying time. This can be eliminated by increasing drying temperature or increasing drying time. The element of interest may be in chemical forms (e.g., MoO_2 and MoO_3). The use of a chemical modifier may cause the formation of different compounds of an element. Double peaks have been observed in slurry sampling, where it was suggested that the first peak may be produced by analyte desorbing from the surface of slurry particles, and the second from analyte trapped within the particles (18). A higher pyrolysis temperature or longer pyrolysis time and/or higher atomization temperature and shorter atomization time may decompose one of the chemical forms and eliminate the double peaks. There exists the possibility that a degraded or well-used graphite tube could induce two forms of an element. This problem is eliminated by use of a new tube.

How Do I Use Characteristic Mass Values? Characteristic mass, or sensitivity, is the amount of analyte required to give an absorbance of 0.0044 s [Section 3.2, Eq. (3.5)]. The characteristic mass is used to give information about the slope of the calibration graph and how well the instrument has been optimized. Generally it should be possible to experimentally obtain sensitivity values within a factor of 2 of the characteristic mass specified by the manufacturer. Experimental measurement of the characteristic mass involves preparation of a calibration graph and use of Eq. (3.5). In general, day-to-day variation of the characteristic mass (19) should be less than 20%.

How Do I Use Limits of Detection? Limits of detection refer to the lowest mass of an element that can be detected quantitatively; usually this means detection with 95% certainty [Section 3.2, Eq. (3.6)]. The limit of detection is measured experimentally by construction of a calibration graph, measurement of the blank 16 times to obtain the standard deviation of the measurement, multiplication of the standard deviation by 3, and substitution into Eq. (3.6). Generally analytical measurements cannot be made near the detection limit because of the degradation in precision. Normally one can expect to do routine analysis a factor of 5 to 10 times above the detection limit. Many analysts use characteristic masses, rather than limits of detection, to evaluate instrument performance because they are easier to measure.

How Is the Spectral Bandpass Selected? Values of the bandpass recommended by the manufacturer are good starting points and probably adequate for many analyses (20). The use of a very wide bandpass may allow more than one HCL line to reach the detector, resulting in large curvature in the calibration graph and a reduced LDR. On the other hand, the use of a very narrow bandpass reduces the amount of light that reaches the detector, which may require the use of high lamp currents to obtain reasonable throughput. Many manufacturers provide emission spectra of hollow cathode lamps that may be used to help select the spectral bandpass as well. If in doubt, optimization may be performed by construction of calibration graphs at various spectral bandpass values.

Is It Necessary to Calibrate in the Linear Range of the Calibration Graph? The simple answer is no, although degradation in precision may be expected (20). Modern instruments employ microcomputers for data collection and manipulation, and most can quantify nonlinear graphs. Some instruments have correction functions that involve manual entry of curvature correction values to optimize the graph. Most modern instruments incorporate automatic linearization algorithms. Recent linearization algorithms have been developed that involve correction for stray light (21–26). In general, one should expect the linear range of GFAAS to be 1 to 3 orders of magnitude, depending on the element, lamp current, spectral bandwidth, analytical wavelength, and method of background correction. Very low LDRs may indicate further optimization of these conditions is required. Some methods of background correction are less suitable for some elements. For example, the LDR of involatile elements (e.g., aluminum, platinum) is typically degraded by the use of self-reversal background correction (Section 4.6.3) (27), and one of the other methods may be preferable.

Can Analyses Be Performed in Organic Solvents? Yes, which is a significant advantage of GFAAS compared to flame or plasma methods (Section 6.1). The solvent is removed in the dry step, eliminating interferences. Organometallic standards may be used to prepare a calibration graph for these analyses.

Are There Particularly Troublesome Elements? Although the difficulty of determining most elements has diminished due to the use of modern furnace technology, two major classes of difficult elements remain: extremely volatile and extremely involatile elements. The use of palladium as a chemical modifier has allowed the use of pyrolysis temperatures of at least 1000°C for most volatile elements, which allows removal of the majority of the matrix. Two exceptions are cadmium and mercury. In spite of this limitation, cadmium has been commonly and accurately determined by GFAAS. In our opinion, mercury is probably easier to determine by other techniques than GFAAS, such as cold vapor AAS. Refractory elements, such as tungsten, boron, silicon, titanium, vanadium, molybdenum, platinum, barium, and the lanthanide elements, are also difficult to determine. Pyrolytically coated graphite is required for these elements, but memory effects may remain even with high atomization temperatures. Platform atomization usually cannot be employed because of their involatility. Transversely heated tubes are more isothermal than longitudinally heated tubes, which may reduce condensation of these elements in cool regions. The addition of modifiers to increase the volatility of these elements may assist in their determination. Barium is particularly difficult to determine because its resonance wavelength is in the visible region (553.5 nm), resulting in high levels of blackbody emission from the tube.

Are There Samples That Are Difficult to Analyze by GFAAS? Some types of sample are regarded as "difficult" by GFAAS. These include samples with high salt concentrations, such as blood, urine, and seawater. Again, the use of modern furnace technology has made determination of many elements in these samples routine. Samples with high silica concentrations (e.g., geological) are difficult to do by GFAAS without removal of the matrix, and may be more easily analyzed by an alternative technique (e.g., inductively coupled plasma emission). Some samples are difficult to dissolve; GFAAS has the advantage that these materials may be analyzed as slurries (Section 6.3.5) or as solids (Section 6.3.6).

What Precision Values Should Be Expected? The continuous determination of a solution should give a precision of 1 to 5%, depending on the concentration level and sample matrix. Typically precision on the order of 2 to 5% RSD can be obtained with routine analysis. Under carefully optimized conditions, it is often possible to obtain better precision than this. The introduction of solids or slurries may degrade the precision depending on the homogeneity of the materials and the reproducibility of the solid sample introduction. Precision values between 5 and 10% are commonly obtained by these methods, although with homogeneous materials the precision may be comparable to values obtained with dissolution methods.

REFERENCES

1. J. Versieck and L. Vanballenberghe, "Collection, Transport, and Storage of Biological Samples for the Determination of Trace Metals," in H. G. Seiler, A. Sigel, and H. Sigel, Eds., *Handbook on Metals in Clinical and Analytical Chemistry*, Marcel Dekker, New York, 1994.
2. A. Aitio, J. Järvisalo, and M. Stoeppler, "Sampling and Sample Storage," in R. F. M. Herber and M. Stoeppler, Eds., *Trace Element Analysis in Biological Specimens*, Elsevier, Amsterdam, 1994.
3. J. R. W. Woittiez and J. E. Sloof, "Sampling and Sample Preparation," in Z. B. Alfassi, Ed., *Determination of Trace Elements*, VCH, Weinheim, 1994.
4. B. Markert, Ed., *Environmental Sampling for Trace Analysis*, VCH, Weinheim, Germany, 1994.
5. B. Markert, *Instrument Element Analysis and Multielement Analysis of Plant Samples*, Wiley, New York, 1996.
6. H. Haraguchi and T. Akagi, "Application of Atomic Absorption Spectrometry to Marine Analysis," in S. J. Haswell, Ed., *Atomic Absorption Spectrometry: Theory, Design, and Application*, Elsevier, Amsterdam, 1991.
7. A. Ashton and R. Chan, *Analyst*, **112**, 841 (1987).

8. E. Prichard, G. M. MacKay, and J. Points, Eds., *Trace Analysis: A Structured Approach to Obtaining Reliable Results*, Royal Society of Chemistry, Cambridge, UK, 1996.
9. G. Benoit, K. S. Hunter, and T. F. Rozan, *Anal. Chem.*, **69**, 1006 (1997).
10. W. P. Robarge and I. Fernandez, *Quality Assurance Methods Manual for Laboratory Analytical Techniques*, U.S. EPA and USDA Forest Program, Corvallis, OR, unpublished.
11. J. M. Christensen, O. M. Poulsen, and T. Anglov, "Method Evaluation, Quality Control, and External Quality Assurance Systems of Analytical Procedures," in H. G. Seiler, A. Sigel, and H. Sigel, Eds., *Handbook on Metals in Clinical and Analytical Chemistry*, Marcel Dekker, New York, 1994.
12. J. Kane, *Spectroscopy*, **9**(5), 18 (1994).
13. R. M. Parr and M. Stoeppler, "Reference Materials for Trace Element Analysis," in R. F. M. Herber and M. Stoeppler, Eds., *Trace Element Analysis in Biological Specimens*, Elsevier, Amsterdam, 1994.
14. R. L. Watters, *Spectrochim. Acta, Part B*, **46B**, 1593 (1991).
15. M. Parkany, Ed., *The Use of Recovery Factors in Trace Analysis*, Royal Society of Chemistry, Cambridge, UK, 1996.
16. B. Welz, *Fresenius J. Anal. Chem.*, **325**, 95 (1986).
17. W. B. Barnett, W. Bohler, G. R. Carnrick, and W. Slavin, *Spectrochim. Acta, Part B*, **40B**, 1689 (1985).
18. H. Qiao and K. W. Jackson, *Spectrochim. Acta, Part B*, **47B**, 1267 (1992).
19. W. Slavin, "Graphite Furnace AAS," in R. F. M. Herber and M. Stoeppler, Eds., *Trace Element Analysis in Biological Specimens*, Elsevier, Amsterdam, 1994.
20. S. J. Haswell, Ed., *Atomic Absorption Spectrometry: Theory, Design, and Applications*, Analytical Spectroscopy Library Ed., Vol. 5, Elsevier, Amsterdam, 1991.
21. B. V. L'vov, L. K. Polzik, N. V. Kocharova, Y. A. Nemets, and A. V. Novichikhin, *Spectrochim. Acta, Part B*, **47B**, 1187 (1992).
22. B. V. L'vov, L. K. Polzik, P. F. Fedorov, and W. Slavin, *Spectrochim. Acta, Part B*, **47B**, 1411 (1992).
23. R. F. Lonardo, A. I. Yuzefovsky, J. X. Zhou, J. T. McCaffrey, and R. G. Michel, *Spectrochim. Acta, Part B*, **51B**, 1309 (1996).
24. E. G. Su, A. I. Yuzefovsky, R. G. Michel, J. T. McCaffrey, and W. Slavin, *Microchem. J.*, **48**, 278 (1993).
25. E. G. Su, A. I. Yuzefovsky, R. G. Michel, J. T. McCaffrey, and W. Slavin, *Spectrochim. Acta, Part B*, **49B**, 367 (1994).
26. A. I. Yuzefovsky, R. F. Lonardo, J. X. Zhou, R. G. Michel, and I. Koltracht, *Spectrochim. Acta, Part B*, **51B**, 713 (1996).
27. C. E. Lee, Determination of Aluminum, Calcium, and Magnesium in Fraser Fir (*Abies fraseri*) Foliage and Surrounding Soil by Atomic Absorption Spectrometry, M.S. Thesis, Western Carolina University, Cullowhee, NC, 1995.
28. *AA Scan-1 Operators Manual*, Thermo Jarrell Ash, Franklin, MA, 1993.

CHAPTER 8

COMMERCIAL GFAAS INSTRUMENTATION: TYPES, COSTS, AND TRAINING

This chapter provides the reader with the basic types and costs of commercial instrumentation, including major accessories. In addition to capital costs associated with the purchase of an instrument, the costs of consumables, service, and maintenance are also discussed. Lastly, options for training on the use of GFAAS instruments and methods are described.

8.1 TYPES AND COSTS OF GFAAS INSTRUMENTATION

Several manufacturers offer commercial GFAAS instrumentation in North America. A summary of these companies and their products is provided in Table 8.1. Several offer two or more instruments, and hence these options are listed. Most instruments are controlled by an IBM-compatible computer, which also functions to process data. Graphite furnace atomic absorption spectrometry software is currently primarily run out of Microsoft Windows, which is menu-driven and considered user-friendly. Several accessories are available with most of these instruments, which include electrodeless discharge lamp power supplies and autosamplers. In addition, there are a few accessories unique to one or two manufacturers, such as a capacitively coupled plasma atomic emission system, a CCD camera to record dry and pyrolysis steps, an automated furnace alignment system, an automated slurry sampling system, or a flow injection unit. At the current time, most manufacturers employ longitudinal heating of the graphite tube, which is the "traditional" method. It is expected that most manufacturers will offer transversely heated systems in the near future. Most manufacturers offer two background correction systems, although a few offer only one method. Several instruments are now available that allow simultaneous multielemental analysis. One instrument is designed for the determination of five specific elements, and a portable instrument is available for the determination of lead.

The cost of GFAAS instrumentation may range from $25,000 for a single element system without an autosampler to $100,000 for a multielemental flame/furnace unit with two or more background correction systems and other accessories. In addition to the instrument manufacturers, accessories and

Table 8.1 Commercial GFAAS Instrumentation

Company	Model	Furnace Heating	Background Correction	Multielemental	Software System	Other
Aurora Instruments (Vancouver, BC)	AI 1000/2000	Transverse	Continuum source	No	Microsoft Windows	Capacitively coupled plasma available
Buck Scientific (East Norwalk, CT)	210 GVP	Longitudinal	Self-reversal, continuum source	No	On-board computer	
Exeter Analytical		Longitudinal	Self-reversal, continuum source	No	Microsoft Windows	Pb only
GBC Scientific Instruments (Arlington Heights, IL)	932 AA, Avanta, Avanta Σ	Longitudinal	Continuum source	No	Microsoft Windows	
Hitachi (San Jose, CA)	Z-5000	Longitudinal	dc Zeeman	4 elements	Microsoft Windows	
Leeman Labs (Lowell, MA)	Analyte-5	Transverse	Self-reversal	6 elements	Microsoft Windows	Designed for simultaneous determination of As, Pb, Se, Sb, and Tl; other elements may be specified at purchase
	Analyte 1	Tungsten filament	Nearby line	No	Microsoft Windows	Portable for Pb only

Perkin-Elmer (Norwalk, CT)	AAnalyst 100	Longitudinal	Continuum source	No	On-board computer	Automated slurry sampling system; flow injection system
	AAnalyst 300	Longitudinal	Continuum source	No	Microsoft Windows	
	4110 ZL	Transverse	Longitudinal ac Zeeman	No	Microsoft Windows	
	SIMAA 6000	Transverse	Longitudinal ac Zeeman	6 elements	Microsoft Windows	
Shimadzu (Columbia, MD)	AA-6601	Longitudinal	Self-reversal, continuum source	No	Microsoft Windows	Automated system available to align furnace
	AA-6701	Longitudinal	Self-reversal, continuum source	No	Microsoft Windows	
Thermo Jarrell Ash-Unicam (Franklin, MA)	989	Longitudinal	Transverse ac Zeeman, continuum source	No	Microsoft Windows	Optional video camera allows observation of tube during dry and pyrolysis steps
Varian (Palo Alto, CA)	SpectrAA 50/55	Longitudinal	Continuum source	No	On-board computer	
	SpectrAA 100/200	Longitudinal	Continuum source	No	Microsoft Windows	
	SpectrAA 880	Longitudinal	Continuum source	No	Microsoft Windows	
	SpectrAA 880Z	Longitudinal	Transverse ac Zeeman	No	Microsoft Windows	

consumables are available from a variety of other suppliers. A list of these companies is provided in the annual *Analytical Chemistry Labguide* and *American Laboratory Buyers Guide*. They provide the accessories at a (usually) reduced cost compared to the manufacturer, although these items may not meet specifications of the manufacturer.

8.2 CONSUMABLES

For best performance, an HCL, costing $150 to $400 each, is required for each element to be determined, and hence a laboratory that is required to determine many elements requires a large inventory of lamps (Section 4.1.1). Multielement lamps are available with some reduction in cost, but the performance may be significantly degraded compared to single-element lamps. A further factor is that lamps do not have an infinite lifetime or shelf life. Typically, HCLs are guaranteed for 2000 mA hours, which means if they are used at 10 mA, then they are guaranteed for 200 h of use, although some lamps last for longer periods. Lamps of volatile elements such as cadmium, selenium, zinc, and so forth need frequent replacement. Also, HCLs will frequently not work if they have not been used over a long period of time.

Electrodeless discharge lamps are recommended for some volatile elements such as arsenic, lead, selenium, cadmium, and so forth. They cost slightly more than an HCL and require an additional power supply, typically costing $2000 to $3000.

For GFAAS, a graphite tube will last around 200 to 500 firings depending on the atomization temperature and the type of sample, dissolution reagents, and chemical modifier. Graphite tubes cost between $20 and $50 each, with the cost determined on the quantity bought, the complexity of tube (e.g., a simple longitudinally heated tube is less expensive than a transversely heated tube with integrated contacts), and the vendor. Platforms have similar lifetime as graphite tubes and cost $10 to $15 based on similar criteria. Platforms are typically integrated into the tube by the manufacturer. Other replacement parts for a furnace, such as graphite electrodes, should be changed every 3 to 12 months depending on amount of use, typically costing $1000 to $2000 per year. A tank of argon, which serves as the purge gas, will generally allow several days of analysis.

These consumables are available from instrument manufacturers and various other suppliers (Section 8.1).

8.3 SERVICE AND MAINTENANCE

Typically, vendors of GFAAS instrumentation provide one year of comprehensive service (parts, travel, and labor) subsequent to installation. After the warranty has expired, the user has the option of paying for service as needed or purchasing a service contract. A full-service contract covers all service calls during the period of the agreement, including labor, travel, and materials, and usually guarantees a service engineer will be on-site within a specified period of time. It typically does not include consumables (Section 8.2) or customer training (Section 8.4). Typical cost for a service contract is approximately 10% of the purchase price of the instrument ($3000–$10,000). A lower cost may be obtained if several instruments (not necessarily atomic absorption instruments) are under contract by the same manufacturer or if a reduced service contract is purchased. Reduced service contracts may have a limited number of service calls during the warranty period, include either parts or labor, or have a limit on the cost of parts or labor provided.

The decision whether or not to purchase a service contract is dependent on the frequency of use and needs of the laboratory. Laboratories that rely on an instrument to be operational 8+ hours a day, 5 days a week are probably best served by a service contract. These instruments are more likely to need service, justifying the cost of the contract. The contract also guarantees that a service engineer will be on-site within a few days to repair the instrument. On the other hand, if an instrument is used on an irregular basis, the need for a service contract diminishes. It should be pointed out, however, that if major repairs are required, a one-year contract will typically pay for itself with one on-site visit.

8.4 TRAINING

During the installation of GFAAS instrumentation, the service engineers of most commercial vendors spend several hours to a day providing rudimentary training on the instrument. In addition, most companies offer 1- to 4-day training courses on the operation of their instrumentation. Frequently these courses are included in the cost of the system and are typically held every few months in the company's headquarters or a major city.

Useful $\frac{1}{2}$- to 3-day courses are also offered on GFAAS at national meetings, which cost from $150 to $500. These courses provide background on GFAAS but are not specific to one company's products.

In the opinion of the authors, GFAAS training requires 1 week to 1 month with an unfamiliar instrument. This period includes familiarity with the operation, routine maintenance, and method development.

CHAPTER 9

FUTURE OF GFAAS

Graphite furnace atomic absorption spectrometry has developed into a mature technique that has been employed for thousands of applications of real sample analysis. The development of multielemental GFAAS instrumentation allows the potential for more rapid analysis by this technique (Section 4.8.2). Recent improvements in instrumentation and methodology have demonstrated the potential of solids analysis without a dissolution step using slurry or solid sampling (Sections 6.3.5 and 6.3.6). Fundamental GFAAS studies continue to examine the chemical and physical processes involved in electrothermal atomization in order to characterize and eliminate interferences (Sections 2.9, 5.1, and 5.2). These developments have ensured that GFAAS will continue to be widely used for practical elemental analysis in the future.

Throughout this volume, we have discussed innovations in GFAAS. We would like to suggest two recent developments in GFAAS instrumentation that may have practical implications for elemental analysis over the next decade or so.

First, laser diodes offer considerable potential for GFAAS (Section 4.1.4), particularly with the recent commercial development of laser diodes in the blue-green (470–515 nm) and blue (420 nm) spectral regions (1). Their tunability and narrow linewidth allow high sensitivity analysis, as well as the potential for interference studies which are not possible with fixed wavelength hollow cathode lamps and electrodeless discharge lamps. Obviously, the integration of laser diodes in commercial GFAAS instrumentation is dependent upon the development of light sources that cover the spectral region between 200 and 400 nm.

Second, portable GFAAS instrumentation (Section 4.6.1) provides elemental analysis capabilities in the field (2). This system may reduce contamination caused by storage and transport of samples to the laboratory for analysis. Future method development with this instrument will determine its suitability for real sample analysis.

Since the first electrothermal atomizer was described by L'vov (3–5), in 1959, GFAAS has been widely employed for elemental analysis. We expect this to continue as further development of this technique continues.

REFERENCES

1. K. Niemax, A. Zybin, C. Schnürer-Patschan, and H. Groll, *Anal. Chem.*, **68**, 351A (1996).
2. C. L. Sanford, S. E. Thomas, and B. T. Jones, *Appl. Spectrosc.*, **50**, 174 (1996).
3. B. V. L'vov, *Ing. Fiz. Zh.*, **2**, 44 (1959).
4. B. V. L'vov, *Ing. Fiz. Zh.*, **2**, 56 (1959).
5. B. V. L'vov, *Spectrochim. Acta*, **17**, 761 (1961).

APPENDIX A

HISTORICAL BACKGROUND

A.1 EARLY WORK IN ATOMIC SPECTROSCOPY (BEFORE 1955)

The visible spectrum was observed in 1666 by Isaac Newton (1). In 1802 Wollaston (2) reported that the sun's spectrum contained a number of dark lines or bands, and in 1817 Fraunhofer (3) measured the wavelengths of 590 of these solar lines. In 1859 Kirchhoff and Bunsen (4) determined that solar lines were caused by the absorption of atomic vapors and recognized the usefulness of atomic absorption for qualitative analysis. Wood (5) used a flame, saturated with sodium as a light source, to observe absorption of light by sodium atoms in an evacuated glass tube. Malinowski (6) reported a linear relationship between concentration of gaseous mercury and absorbance in 1914. Paschen (7) developed rudimentary hollow cathode lamps in 1916, although they were not employed with flames until 40 years later. Ballard and Thornton (8) determined mercury in air by an atomic absorption procedure in the 1940s.

A.2 FLAME ATOMIC ABSORPTION SPECTROMETRY

In 1955, Walsh (9) and Alkemade and Milatz (10,11) independently published the first papers that described the use of a flame as an atom cell for atomic absorption. Many features of modern flame AAS (12,13) instrumentation were incorporated in Walsh's original paper, including the use of hollow cathode lamps as light sources, modulation to discriminate against flame emission, and a long-path burner. Other significant developments in flame AAS include the production of the first commercial instrument by Perkin-Elmer in 1961; the development of the nitrous oxide/acetylene flame by Amos and Willis (14), which allowed determination of involatile and refractory elements (e.g., aluminum, chromium); and the introduction of continuum source background correction by Koirtyohann and Pickett (15,16) (Section 4.6.2).

A.3 GRAPHITE FURNACE ATOMIC ABSORPTION SPECTROMETRY

The graphite furnace, which is also called the electrothermal atomizer (ETA), was commercially developed for AAS in the mid-1960s. Its principal advantages compared to flame atomizers include an improvement in detection limits, typically 10 to 100 times; the ability to introduce microvolume (2–20 μL) and micromass (1–5 mg with direct introduction of solids) samples; and the ability to thermally pretreat samples using a pyrolysis step before atomization of the analyte. However, early graphite furnace designs were prone to interferences, and a considerable research was required to develop instruments that were capable of routine analysis.

The use of an electrically heated tubular furnace was first reported by King in 1905 (17), but the instrument developed by L'vov in the late 1950s (18–20) is regarded as the forerunner of present-day ETAs. The system consisted of a graphite electrode, onto which the sample was applied, and a graphite tube, each with a separate power supply. The sample was atomized from the electrode into the preheated tube and was similar in design to a modern two-step furnace (Fig. 4.14). The system was difficult to operate and the precision was poor.

In 1968, Massmann (21) described a heated graphite atomizer (HGA), which was later commercially developed by the Perkin-Elmer Corporation (Norwalk, CT) and proved to be the forerunner for all current commercial ETAs. The Massmann system used a graphite tube of dimensions 50 mm long and 10 mm diameter that was heated by electrical resistance, typically 7 to 10 V at 400 A. An inert gas, usually argon or nitrogen, flowed at a constant rate of around 1.5 L/min, and the electrodes were water-cooled. The sample (5–100 μL) was deposited through a dosing hole in the center of the tube. The tube was heated in dry, pyrolysis, and atomization steps by variation of the applied voltage. Careful control of the temperature was required in order to obtain good reproducibility.

Woodriff (22,23) constructed a constant-temperature graphite furnace that was maintained at 1500 to 2500°C. Samples were introduced continuously, which improved the precision compared to the transient signals obtained with other systems. This furnace design was only briefly developed commercially, although recent work has shown the advantages of atomization under isothermal conditions.

In 1969 West and co-workers (24) developed a rod or filament atomizer. It consisted of a graphite filament 40 mm in length and 2 mm in diameter, supported by water-cooled electrodes, that could be heated rapidly by the use of current of 70 A at 10 to 12 V. The filament was shielded from the air by a stream of inert gas. The West filament was the forerunner for the mini-Massmann atomizer developed commercially by Varian Associates (Palo Alto, CA) called the carbon rod atomizer (CRA 63 and CRA 90). Advantages included the simpler design compared to the HGA, lower power requirements, and faster (~1000°C/s) heating rate. A small resistively heated graphite cup, which was held between two spring-loaded graphite rods, served as the atom cell. The system was proposed for low microliter volumes, typically 1 to 20 μL. In general, degraded detection limits, particularly for involatile elements, and increased interferences were found using the CRA-type system compared to the HGA because atomization occurred into the relatively cool volume above the cup. Consequently, commercial production of the graphite cup atomizer ended in the mid-1980s.

In the late 1970s and 1980s, a number of instrumental developments improved the precision and accuracy of graphite furnace methods (25–29). These developments, which together are referred to as modern furnace technology (Table 1.1), include the use of platform or probe atomization to ensure atomization occurs under isothermal conditions, integrated absorbance, pyrolytically coated graphite tubes, an autosampler, fast heating rates during the atomization step, fast electronics, chemical modifiers, and modern methods of background correction (Zeeman effect and self-reversal). The ease of use and analytical performance of GFAAS were significantly improved by modern furnace technology. In the late 1980s and early 1990s, the use of transversely heated graphite furnaces (transversely heated graphite atomizers, THGAs) was shown to reduce temperature gradients in the tube and further reduce interferences (30,31). In the mid-1990s, fast furnace programs (32–37) and multielemental GFAAS (38–44) were major foci to increase the sample throughput of the technique (Sections 4.2.2 and 4.8.2, respectively).

More detailed information regarding the historical development of GFAAS is given in the references (1,10,12,25,26,29,45–48).

REFERENCES

1. C. T. J. Alkemade and R. Herrmann, *Fundamentals of Analytical Flame Spectroscopy*, Wiley, New York, 1979.
2. W. H. Wollaston, *Phil. Trans. R. Soc.*, **A92**, 365 (1802).
3. J. Fraunhofer, *Ann. Phys. (Gilberts Ann.)*, **56**, 264 (1817).
4. G. R. Kirchhoff and R. Bunsen, *Pogg. Ann. Phys. Chem.*, **110**, 161 (1860).
5. R. W. Wood, *Phil. Mag.*, **10**, 513 (1905).
6. A. V. Malinowski, *Ann. Phys.*, **44**, 935 (1914).
7. F. Paschen, *Ann. Phys.*, **50**, 901 (1916).
8. A. E. Ballard and C. D. W. Thornton, *Ind. Engng. Chem., Anal. Edn.*, **17**, 893 (1941).
9. A. Walsh, *Spectrochim. Acta*, **7**, 108 (1955).
10. C. T. J. Alkemade, *J. Opt. Soc. Am.*, **45**, 583 (1955).
11. C. T. J. Alkemade and J. M. W. Milatz, *Appl. Sci. Res.*, **B4**, 289 (1955).
12. J. W. Robinson, *Anal. Chem.*, **66**, 472A (1994).
13. A. Walsh, *Anal. Chem.*, **63**, 933A (1991).
14. M. D. Amos and J. B. Willis, *Spectrochim. Acta*, **22**, 1325 (1963).
15. S. R. Koirtyohann and E. E. Pickett, *Anal. Chem.*, **37**, 601 (1965).

16. S. R. Koirtyohann and E. E. Pickett, *Anal. Chem.*, **38**, 585 (1966).
17. A. S. King, *Astrophys. J.*, **21**, 236 (1905).
18. B. V. L'vov, *Ing. Fiz. Zh.*, **2**, 44 (1959).
19. B. V. L'vov, *Ing. Fiz. Zh.*, **2**, 56 (1959).
20. B. V. L'vov, *Spectrochim. Acta*, **17**, 761 (1961).
21. H. Massmann, *Spectrochim. Acta, Part B*, **23B**, 215 (1968).
22. R. Woodriff, *Appl. Spectrosc.*, **22**, 408 (1968).
23. R. Woodriff, *Appl. Spectrosc.*, **28**, 413 (1974).
24. T. S. West and X. K. Williams, *Anal. Chim. Acta*, **45**, 27 (1969).
25. W. Slavin, *Graphite Furnace AAS: A Source Book*, Perkin-Elmer Corporation, Norwalk, CT, 1984.
26. B. Welz, *Atomic Absorption Spectrometry*, 2nd ed., VCH, Weinheim, Germany, 1985.
27. S. J. Haswell, Ed., *Atomic Absorption Spectrometry: Theory, Design, and Applications*, Analytical Spectroscopy Library Ed., Vol. 5, Elsevier, Amsterdam, 1991.
28. L. H. J. Lajunen, *Spectrochemical Analysis by Atomic Absorption and Emission*, Royal Society of Chemistry, Cambridge, England, 1992.
29. B. V. L'vov, *Anal. Chem.*, **63**, 924A (1991).
30. W. Frech, D. C. Baxter, and B. Hütsch, *Anal. Chem.*, **58**, 1973 (1986).
31. M. Sperling, B. Welz, J. Hertzberg, C. Rieck, and G. Marowsky, *Spectrochim. Acta, Part B*, **51B**, 897 (1996).
32. D. J. Halls, *J. Anal. Atom. Spectrom.*, **10**, 169 (1995).
33. M. Hoenig and A. Cilissen, *Spectrochim. Acta, Part B*, **48B**, 1003 (1993).
34. L. Lian, *Spectrochim. Acta, Part B*, **47B**, 239 (1992).
35. V. A. Ganadillo, L. P. de Machado, and R. A. Romero, *Anal. Chem.*, **66**, 3624 (1994).
36. P. J. Parsons and W. Slavin, *Spectrochim. Acta, Part B*, **48B**, 925 (1993).
37. Z. Li, G. Carnrick, and W. Slavin, *Spectrochim. Acta, Part B*, **48B**, 1435 (1993).
38. B. Radziuk, G. Rödel, H. Stenz, H. Becker-Ross, and S. Florek, *J. Anal. Atom. Spectrom.*, **10**, 127 (1995).
39. B. Radziuk, G. Rödel, M. Zeiher, S. Mizuno, and K. Yamamoto, *J. Anal. Atom. Spectrom.*, **10**, 415 (1995).
40. M. Berglund, W. Frech, and D. C. Baxter, *Spectrochim. Acta, Part B*, **46B**, 1767 (1991).
41. J. G. Sen Gupta, *Talanta*, **40**, 791 (1993).
42. J. G. Sen Gupta, *J. Anal. Atom. Spectrom.*, **8**, 93 (1993).
43. D. R. Demers and M. C. Almeida, *Am. Environ. Lab.*, **June**, 13 (1995).
44. K. S. Farah and J. Sneddon, *Appl. Spectrosc. Rev.*, **30**, 351 (1995).
45. W. Slavin and G. R. Carnrick, *CRC Crit. Rev. Anal. Chem.*, **19**, 95 (1988).
46. B. V. L'vov, *J. Anal. Atom. Spectrom.*, **3**, 9 (1988).
47. B. V. L'vov, *Spectrochim. Acta, Part B*, **39B**, 149 (1984).
48. H. Falk, *CRC Crit. Rev. Anal. Chem.*, **19**, 29 (1988).

APPENDIX B

GFAAS LITERATURE

Graphite furnace atomic absorption is a widely used technique, with several hundred articles published per year, including fundamental studies and applications (1). In addition, several books and reviews appear annually, including reviews on specific topics on atomic absorption that are discussed in the appropriate section of this volume. This section is not intended to be comprehensive but rather to provide the reader with the most relevant, recent literature.

B.1 BOOKS

Several recent volumes focus on GFAAS, separately, or with flame AAS, or with other atomic spectrometry methods. Welz (2) discusses flame and furnace atomic absorption, including sections on instrumentation, methodology, discussion of individual elements, and specific applications. Haswell (3) has edited a volume that includes flame and furnace atomic absorption instrumentation, practical methodology, and applications. Varma (4) focuses on the development of protocols for the determination of elements by GFAAS and includes standardized conditions for individual elements. Lajunen (5) describes various atomic spectrometry techniques, including atomic absorption. Metcalfe (6) has written a volume in the *Analytical Chemistry by Open Learning* series, which is designed for training and continuing education of technical staff, on atomic absorption and atomic emission. Robinson (7) and Heckmann and Trabert (8) discuss theory and instrumentation of all forms of atomic spectrometry. Minoia and Caroli (9) have edited a volume that focuses on Zeeman background correction for GFAAS and its application to environmental and biological samples. Slavin (10) discusses GFAAS instrumentation, the vaporization process, applications, and the analysis of individual elements.

A number of volumes have been published that focus on elemental analysis and include GFAAS. Seiler et al. (11) have edited a volume that describes sampling, sample preparation, and various methods of elemental analysis. Chapters are included on each element that describe that element's chemistry, distribution, toxicology, physiology, and analytical determination. A volume edited by Alfassi (12) focuses on various methods of elemental analysis. Taylor et al. (13) have written a book whose principal goal is to help an analyst choose a method of elemental analysis. Vandecasteele and Block (14) describe various methods of elemental analysis, including GFAAS, with an emphasis on environmental and biological applications. Potts (15) has written a volume that focuses on silicate rock analysis but that also includes detailed descriptions of methods of elemental analysis. Herber and Stoeppler (16) describe the various methods of elemental analysis for biological samples. Smoley (17) has published a volume listing United States Environmental Protection Agency (USEPA) methods for the determination of metals in environmental samples by a variety of techniques. This volume is valuable for laboratories interested in monitoring elements according to USEPA specifications. Markert (18) describes instrumental methods of analysis for plant samples. Sneddon (19,20) edits a series of volumes that contain review articles on atomic spectrometry.

B.2 PERIODICALS

Atomic absorption publications appear in a variety of analytical chemistry journals, including *Analyst*, *Analytica Chimica Acta*, *Analytical Chemistry*, *Analytical Letters*, *Analytical Proceedings*, *Applied Spectroscopy*, *Fresenius Journal of Analytical Chemistry*, *Microchemical Journal*, *Mikrochimica Acta*, *Spectroscopy Letters*, and *Talanta*. Two instrument manufacturers, Perkin-Elmer and Thermo Jarrell Ash, publish *Atomic Spectroscopy* and *The Spectroscopist*, respectively. Two periodicals focus on elemental analysis: *Journal of Analytical Atomic Spectrometry* and *Spectrochimica Acta, Part B*, and, in addition to contributed articles, often have invited articles and conference proceedings on atomic absorption. *CRC Critical Reviews in Analytical Chemistry* includes comprehensive review articles on GFAAS. *Spectrochimica Acta Reviews* (previously called *Progress in Analytical Spectrometry* and *Progress in Analytical Atomic Spectrometry*), which was incorporated into *Spectrochimica Acta, Part B* in 1994, also contains comprehensive reviews on atomic absorption.

Readers interested in applications of ETA-AAS in a specific discipline are referred to articles in that area. For example, toxicological applications are published in *Journal of Analytical Toxicology*, clinical applications in *Clinical Chemistry*, soil analyses in *Soil Science Society of America Journal*, and pharmaceutical analysis in *Journal of Pharmaceutical Science*.

B.3 REVIEWS

Analytical Chemistry published a biennial issue on fundamental reviews of analytical techniques, and one article discusses atomic absorption spectrometry

(21–23). *Journal of Analytical Atomic Spectrometry* publishes a series of annual reviews that focus on environmental analysis (24); clinical and biological materials, foods, and beverages (25); advances in atomic absorption and fluorescence instrumentation and methodology (26); and industrial analysis (27). This periodical also includes a comprehensive update of articles published in the area of atomic spectrometry. A database (*JAASbase*) of these references is available from the periodical's publisher, the Royal Society of Chemistry. *Atomic Spectroscopy* also includes an annual list of atomic spectrometry references.

B.4 INTERNET RESOURCES

Many GFAAS Internet resources are currently available, and these were summarized by O'Haver (28). These include courses, software, tutorials, commercial instrument manufacturers, and discussion lists. Wiese (29) published a list of atomic spectroscopic databases on the World Wide Web.

B.5 APPLICATIONS OF GFAAS

Graphite furnace atomic absorption spectroscopy has been employed for thousands of analyses of a variety of samples. The purpose of this section is not to describe all of these applications (which would require several volumes) but rather to list some useful reviews and volumes as starting points in Table B.1. The primary literature is then accessible through these sources.

Table B.1 Summary of Some Reviews and Volumes on GFAAS Applications

Type of Sample	References
Clinical chemistry (tissues and fluids)	3, 9, 11, 16, 25, 30, 31
Environmental science (water, soil, plants)	3, 9, 18, 24, 32–39
Forensic chemistry	3, 25, 31, 40
Foods	3, 9, 25, 41
Metallurgy	3, 27, 42
Air	3, 24, 43–45
Geochemistry	3, 15, 24, 46, 47
Chemicals	3, 27, 48, 49
Petroleum	3, 27, 47
Glasses and ceramics	3, 27
Semiconductors	3, 27
Pharmaceuticals	3, 25

A few general comments can be made about the GFAAS applications literature. An excellent source is the volume by Haswell (3), which has many chapters on specific applications of atomic absorption. Two periodicals regularly publish review articles that include GFAAS applications. *Journal of Analytical Atomic Spectrometry* publishes three annual reviews of applications (24,25,27), and *Analytical Chemistry* publishes a biennial issue entitled Application Reviews that includes GFAAS analyses.

REFERENCES

1. B. V. L'vov and W. Slavin, *J. Anal. Atom. Spectrom.*, **6**, 191 (1991).
2. B. Welz, *Atomic Absorption Spectrometry*, 2nd ed., VCH, Weinheim, Germany, 1985.
3. S. J. Haswell, Ed., *Atomic Absorption Spectrometry: Theory, Design, and Applications*, Analytical Spectroscopy Library Ed., Vol. 5, Elsevier, Amsterdam, 1991.
4. A. Varma, *CRC Handbook of Furnace Atomic Absorption Spectrometry*, CRC Press, Boca Raton, FL, 1990.
5. L. H. J. Lajunen, *Spectrochemical Analysis by Atomic Absorption and Emission*, Royal Society of Chemistry, Cambridge, England, 1992.
6. E. Metcalfe, *Atomic Absorption and Emission Spectroscopy*, Analytical Chemistry by Open Learning, B. R. Currell, Ed., Wiley, New York, 1987.
7. J. W. Robinson, *Atomic Spectroscopy*, Marcel Dekker, New York, 1990.
8. P. H. Heckmann and E. Trabert, *Introduction to the Spectroscopy of Atoms*, Elsevier, Amsterdam, 1989.
9. C. Minoia and S. Caroli, Eds., *Applications of Zeeman Graphite Furnace Atomic Absorption Spectrometry in the Chemical Laboratory and Toxicology*, Pergamon, Oxford, UK, 1992.
10. W. Slavin, *Graphite Furnace AAS: A Source Book*, Perkin-Elmer Corporation, Norwalk, CT, 1984.
11. H. G. Seiler, A. Sigel, and H. Sigel, Eds., *Handbook on Metals in Clinical and Analytical Chemistry*, Marcel Dekker, New York, 1994.
12. Z. B. Alfassi, Ed., *Determination of Trace Elements*, VCH, Weinheim, Germany, 1994.
13. L. R. Taylor, R. B. Papp, and B. D. Pollard, *Instrumental Methods for Determining Elements*, VCH, Weinheim, Germany, 1994.
14. C. Vandecasteele and C. B. Block, *Modern Methods for Trace Element Determination*, Wiley, New York, 1993.
15. P. J. Potts, *A Handbook of Silicate Rock Analysis*, Blackie, Glasgow, 1987.
16. R. F. M. Herber and M. Stoeppler, *Trace Element Analysis in Biological Specimens*, Elsevier, Amsterdam, 1994.

17. C. K. Smoley, *Methods for the Determination of Metals in Environmental Samples*, CRC Press, Boca Raton, FL, 1992.
18. B. Markert, *Instrument Element Analysis and Multielement Analysis of Plant Samples*, Wiley, New York, 1996.
19. J. Sneddon, Ed., *Advances in Atomic Spectroscopy*, Vol. 1, Jai Press, Greenwich, CT, 1992.
20. J. Sneddon, Ed., *Advances in Atomic Spectroscopy*, Vol. 2, Jai Press, Greenwich, CT, 1995.
21. K. W. Jackson and H. Qiao, *Anal. Chem.*, **64**, 50R (1992).
22. K. W. Jackson and T. M. Mahmood, *Anal. Chem.*, **66**, 252R (1994).
23. K. W. Jackson and G. Chen, *Anal. Chem.*, **68**, 231R (1996).
24. M. S. Cresser, L. M. Garden, J. A. Armstrong, J. R. Dean, P. J. Watkins, and M. Cave, *J. Anal. Atom. Spectrom.*, **11**, 19R (1996).
25. A. Taylor, S. Branch, H. M. Crews, D. J. Halls, and M. White, *J. Anal. Atom. Spectrom.*, **11**, 103R (1996).
26. S. J. Hill, J. B. Dawson, W. J. Price, I. L. Shuttler, and J. F. Tyson, *J. Anal. Atom. Spectrom.*, **11**, 281R (1996).
27. J. S. Crighton, J. Carroll, B. Fairman, J. Haines, and M. Hinds, *J. Anal. Atom. Spectrom.*, **11**, 461R (1996).
28. T. C. O'Haver, *Spectroscopy*, **11**(1), 12 (1996).
29. W. L. Wiese, *Spectrochim. Acta, Part B*, **52B**, 279 (1997).
30. D. J. Anderson and F. Van Lente, *Anal. Chem.*, **67**, 377R (1995).
31. K. S. Subramanian, *Spectrochim. Acta, Part B*, **51B**, 291 (1996).
32. W. P. Robarge and I. Fernandez, *Quality Assurance Methods Manual for Laboratory Analytical Techniques*, U.S. EPA and USDA Forest Program, Corvallis, OR, unpublished.
33. H. van der Jagt and P. J. Stuyfzand, *Fresenius J. Anal. Chem.*, **354**, 32 (1996).
34. B. Markert, Ed., *Environmental Sampling for Trace Analysis*, VCH, Weinheim, Germany, 1994.
35. A. L. Page, R. H. Miller, and D. R. Keeney, Eds., *Methods of Soil Analysis, Part 2: Chemical and Microbiological Properties*, Vol. 2, American Society of Agronomy and Soil Science Society of America, Madison, WI, 1982.
36. R. E. Clement, G. A. Eiceman, and C. J. Koestler, *Anal. Chem.*, **67**, 221R (1995).
37. P. MacCarthy, R. W. Klusman, S. W. Cowling, and J. A. Rice, *Anal. Chem.*, **67**, 525R (1995).
38. A. E. Greenberg, R. R. Trussell, and L. S. Clesceri, Eds., *Standard Methods for the Examination of Water and Wastewater*, American Public Health Association, Washington, DC, 1992.
39. E. Prichard, G. M. MacKay, and J. Points, Eds., *Trace Analysis: A Structured Approach to Obtaining Reliable Results*, Royal Society of Chemistry, Cambridge, UK, 1996.

40. T. A. Brettell and R. Saferstein, *Anal. Chem.*, **67**, 273R (1995).
41. S. K. C. Chang, E. Holm, J. Schwarz, and P. Rayas-Duarte, *Anal. Chem.*, **67**, 127R (1995).
42. T. R. Dulski, *Anal. Chem.*, **67**, 21R (1995).
43. J. P. Lodge, Ed., *Methods of Air Sampling and Analysis*, Lewis, Chelsea, MI, 1989.
44. D. L. Fox, *Anal. Chem.*, **67**, 183R (1995).
45. J. E. Adkins and N. W. Henry, *Anal. Chem.*, **67**, 349R (1995).
46. L. L. Jackson, P. A. Baedecker, T. L. Fries, and P. J. Lamothe, *Anal. Chem.*, **67**, 71R (1995).
47. J. B. Hooper, *Anal. Chem.*, **67**, 315R (1995).
48. D. G. Anderson, *Anal. Chem.*, **67**, 33R (1995).
49. B. B. Sitholé, *Anal. Chem.*, **67**, 87R (1995).

APPENDIX C

CONDITIONS FOR GFAAS

Table C.1 lists instrumental conditions that may be used as starting points for GFAAS determinations. These conditions have been taken from a variety of sources, and may be regarded as compromise conditions. The instrument manufacturers vary in the recommended wavelengths and bandpasses for some elements, and hence more than one has been listed in these cases. It is normally expected that the type and amount of chemical modifier, pyrolysis temperature, and atomization temperature must be optimized for each type of sample. The limits of detection listed are average values from several sources.

Table C.2 is a list of alternative wavelengths for GFAAS. The relative limit of detection for each line is given, with the primary line given a value of 1 and less sensitive lines given values greater than 1. For some elements (e.g., lead), an alternative line is more sensitive than the primary line and has a relative value less than 1.

Table C.3 is a list of preparation procedures for standards of elements commonly determined by GFAAS. For most analyses, the use of commercially available standards (Analytical Chemistry Labguide) are recommended, but for rarely determined elements these standards should give good results. Preparation of standards may also be performed at a significantly lower cost. Several of these procedures involve the use of metal chlorides or hydrochloric acid, which may interfere with the determination of volatile elements. An alternative for furnace work involves the dissolution of metal nitrates, if available in high purity.

We recommend that for highest accuracy prepared solutions of chemical modifiers be purchased from chemical suppliers (Analytical Chemistry Labguide). However, it is also possible to prepare these solutions at a lower cost using the procedures in Table C.4. The 1% stock solutions are generally diluted to 0.1% for use in analysis.

Table C.1 Conditions for GFAAS (1–6)

Element	Wavelength, nm	Bandpass, nm	Atomization	Recommended Modifier(s)	Pyrolysis Temperature, °C	Atomization Temperature, °C	Detection Limit, pg
Aluminum	309.3 or 396.2	1 or 0.7	Platform	$Mg(NO_3)_2$; $Mg(NO_3)_2 + HCl$	1500	2500	5
Antimony	217.6	0.4–0.7	Platform	$Pd(NO_3)_2 + Mg(NO_3)_2$; $Ni(NO_3)_2$	1000	2400	20
Arsenic	193.7	0.7–1	Platform	$Pd(NO_3)_2 + Mg(NO_3)_2Ni(NO_3)_2$	1100	2300	20
Barium	553.5	0.2–0.4	Wall		1100	2500	10
Beryllium	234.9	0.4–0.7	Platform	$Mg(NO_3)_2$	1100	2400	1
Bismuth	223.1	0.2–0.4	Platform	$Pd(NO_3)_2 + Mg(NO_3)_2$; $Ni(NO_3)_2$	1000	2100	10
Boron	249.7	0.7	Wall	$Ca(NO_3)_2$	1000	2700	1000
Cadmium	228.8	0.7–1	Platform	$(NH_4)_2HPO_4 + Mg(NO_3)_2$; $Pd(NO_3)_2 + Mg(NO_3)_2$	700	1700	0.2
Calcium	422.7	0.7–1	Wall		1100	2500	1
Cesium	852.1	0.7–1	Platform	H_2SO_4	800	1900	5
Chromium	357.9	0.4–0.7	Platform	$Mg(NO_3)_2$	1400	2500	2
Cobalt	242.5 or 240.7	0.2 or 0.4	Platform	$Mg(NO_3)_2$	1200	2500	5
Copper	324.7	0.7–1	Platform	$Pd(NO_3)_2 + Mg(NO_3)_2$	1000	2300	2
Dysprosium	421.2	0.2	Wall		1300	2700	50
Erbium	400.8	0.2–0.4	Wall		1500	2700	50
Europium	459.4	0.2–0.4	Wall		1200	2600	20
Gadolinium	407.9 or 368.4	0.15	Wall		1500	2700	1600
Gallium	294.4 or 287.4	0.7 or 0.4	Platform	$Pd(NO_3)_2 + Mg(NO_3)_2$; $Mg(NO_3)_2$	1000	2200	10

Germanium	265.1	0.15	Platform	$Pd(NO_3)_2 + Mg(NO_3)_2$; $Mg(NO_3)_2$	1000	2400	20
Gold	242.8	0.7–1	Platform	$Pd(NO_3)_2 + Mg(NO_3)_2$; $Ni(NO_3)_2$	1000	2200	10
Indium	325.6 or 303.9	0.7 or 0.15	Platform	$Pd(NO_3)_2 + Mg(NO_3)_2$	1000	2100	5
Iridium	264.0 or 208.8	0.15	Wall		1200	2700	200
Iron	248.3	0.2–0.4	Platform	$Mg(NO_3)_2$	1200	2400	5
Lanthanum	550.1	0.2–0.4	Wall		1400	2700	
Lead	283.3 or 217.0	0.7 or 1	Platform	$(NH_4)_2HPO_4 + Mg(NO_3)_2$; $Pd(NO_3)_2 + Mg(NO_3)_2$; $(NH_4)_2HPO_4$	800	1800	5
Lithium	670.8	0.2–0.4	Platform		800	2600	2
Magnesium	285.2	0.7–1	Platform		900	1800	1
Manganese	279.5	0.2–0.4	Platform	$Mg(NO_3)_2$	1200	2200	1
Mercury	253.7	0.7	Platform	$Pd(NO_3)_2$ pretreated to 1000°C before sample addition	140	2000	50
Molybdenum	313.3	0.4–0.7	Wall		1400	2700	5
Neodynium	463.4 or 492.5	0.15	Wall		1300	2700	1800
Nickel	232.0	0.15	Platform		1200	2500	10
Osmium	290.9	0.2–0.4	Wall		200	2700	300
Palladium	247.6	0.4–0.7	Platform		800	2700	30
Phosphorous	213.6	0.7–1	Platform	$Pd(NO_3)_2 + Ca(NO_3)_2$; $Ni(NO_3)_2$	1200	2700	5000
Platinum	265.9	0.4–0.7	Wall		1100	2700	200
Potassium	766.5	0.4–0.7	Wall		900	2000	1

(continued)

Table C.1 *(continued)*

Element	Wavelength, nm	Bandpass, nm	Atomization	Recommended Modifier(s)	Pyrolysis Temperature, °C	Atomization Temperature, °C	Detection Limit, pg
Rhodium	343.5	0.2–0.4	Wall		1100	2500	10
Rubidium	780.0	0.4–0.7	Platform		700	2100	5
Ruthenium	349.9	0.15	Wall		1200	2500	30
Samarium	429.7	0.2–0.4	Wall		1200	2600	240
Selenium	196.0	1–2	Platform	$Pd(NO_3)_2 + Mg(NO_3)_2$; $Ni(NO_3)_2 + Mg(NO_3)_2$	1000	2300	20
Silicon	251.6	0.2–0.4	Platform	$Ca(NO_3)_2 + La$	1200	2700	30
Silver	328.1	0.7–1	Platform	$Pd(NO_3)_2$; $(NH_4)_2HPO_4$	1000	2000	1
Sodium	589.0	0.4–0.7	Wall		700	1800	1
Strontium	460.7	0.4–0.7	Wall		1100	2600	2
Tellurium	214.3	0.2–0.4	Platform	$Pd(NO_3)_2 + Mg(NO_3)_2Ni(NO_3)_2$	1000	2300	10
Thallium	276.8	0.7–1	Platform	$Pd(NO_3)_2 + Mg(NO_3)_2(NH_4)_2HPO_4$	900	1600	5
Thulium	371.8	0.2	Wall		1500	2700	20
Tin	286.3 or 235.5	0.7 or 0.4	Platform	$Pd(NO_3)_2 + Mg(NO_3)_2(NH_4)_2 HPO_4 + Mg(NO_3)_2$	800	2200	20
Titanium	364.3	0.2–0.4	Wall		1400	2700	40
Uranium	351.5 or 358.5	0.15	Wall		900	2700	10,000
Vanadium	318.5	0.4–0.7	Wall	$Ni(NO_3)_2$	1000	2700	20
Ytterbium	398.8	0.2–0.4	Wall		1200	2500	5
Yttrium	410.2	0.2–0.4	Wall		1200	2700	10,000
Zinc	213.9	0.7–1	Platform	$Ni(NO_3)_2$	600	1800	1

Table C.2 Alternative Absorption Wavelengths for Atomic Absorption Spectrometry (1–3,5–7)[a]

Element	Wavelength, nm	Spectral Bandwidth, nm	Relative Limit of Detection
Aluminum	309.3	1	1
	396.2	0.5	1.2
	308.2	0.5	1.7
	394.4	0.5	2.5
	257.5	0.3	6
	256.8	0.3	15
Antimony	217.6	0.3	1
	206.8	0.5	1
	231.2	0.5	2
	212.7	1	20
Arsenic	193.7	1	1
	189.0	1	1
	197.2	1	2
Barium	553.5	0.5	1
	350.1	0.5	1000
Bismuth	223.1	0.3	1
	222.8	0.3	3
	306.8	0.5	4
	206.2	0.3	8
	227.7	0.5	15
Boron	249.7	1	1
	208.9	1	2
Cadmium	228.8	1	1
	326.1	0.15	400
Calcium	422.7	1	1
	239.9	0.15	100
Cesium	852.1	1	1
	894.5	1	1
	455.5	1	10
	459.3	0.15	20
Chromium	357.9	0.5	1
	359.4	0.5	2
	360.5	0.5	2
	425.4	0.5	3
	520.8	0.15	200
Cobalt	242.5	0.15	1
	240.7	0.3	1
	241.2	0.15	2
	304.4	0.5	10
	352.7	0.15	20

(continued)

Table C.2 *(continued)*

Element	Wavelength, nm	Spectral Bandwidth, nm	Relative Limit of Detection
Copper	324.7	1	1
	327.4	0.5	3
	217.9	0.3	10
	216.5	0.15	20
	249.2	0.5	80
Gallium	287.4	0.3	1
	294.4	0.3	1
	417.2	0.3	1
	250.0	0.3	10
	272.0	0.3	20
Germanium	265.1	0.5	1
	271.0	0.5	2
	259.2	0.5	2
	269.1	0.5	4
	303.9	0.5	20
Gold	242.8	1	1
	267.6	0.5	2
	312.3	1	900
	274.8	1	1100
Indium	303.9	0.15	1
	325.6	0.15	1
	410.5	0.15	3
	256.0	0.15	10
	275.4	0.15	30
Iridium	264.0	0.15	1
	208.8	0.15	0.5
	266.5	0.15	2
	254.4	0.15	3
	351.4	0.15	10
Iron	248.3	0.3	1
	248.8	0.3	2
	271.9	0.3	4
	372.0	0.3	10
	305.9	0.3	30
Lead	283.3	0.5	1
	217.0	1	0.5
	261.4	0.5	30
	202.2	0.5	50
	205.3	0.5	340
Lithium	670.8	0.5	1
	323.3	0.15	500

Table C.2 *(continued)*

Element	Wavelength, nm	Spectral Bandwidth, nm	Relative Limit of Detection
Magnesium	285.2	1	1
	202.6	1	25
Manganese	279.5	0.15	1
	279.8	0.15	2
	280.1	0.15	2
	403.1	0.3	10
Molybdenum	313.3	0.5	1
	317.0	0.5	10
	379.8	0.5	10
	386.4	0.5	20
	311.2	0.3	50
Nickel	232.0	0.15	1
	231.1	0.5	2
	341.5	0.5	5
	303.8	0.5	1
	294.4	0.5	50
Palladium	247.6	0.3	1
	244.8	0.3	2
	276.3	0.3	3
	340.5	1	3
Phosphorous	213.6	0.15	1
	214.9	0.15	2
Platinum	265.9	0.5	1
	306.5	0.5	2
	283.0	0.3	3
	293.0	0.5	4
	271.9	0.3	8
Potassium	766.5	1	1
	769.9	1	3
	404.4	0.3	400
Rhodium	343.5	0.5	1
	369.1	0.5	2
	339.7	0.5	3
Rubidium	780.0	1	1
	794.8	0.3	3
	420.2	0.3	10
	421.6	0.3	30
Ruthenium	349.9	0.15	1
	372.8	0.15	1
	379.9	0.15	2
	392.6	0.15	10

(continued)

Table C.2 *(continued)*

Element	Wavelength, nm	Spectral Bandwidth, nm	Relative Limit of Detection
Selenium	196.0	1	1
	204.0	0.5	10
	206.3	0.5	20
	207.5	0.5	200
Silicon	251.6	0.3	1
	250.7	0.5	2
	252.8	0.3	2
	221.7	0.3	4
	288.2	0.3	20
Silver	328.1	1	1
	338.3	0.5	2
Sodium	589.0	0.5	1
	589.6	1	20
Tellurium	214.3	0.5	1
	225.9	0.5	10
	238.6	0.3	100
Thallium	276.8	1	1
	377.6	1	3
	238.0	0.15	20
	258.0	0.15	20
Tin	286.3	0.5	1
	224.6	0.15	0.5
	235.5	0.5	1
	254.7	0.5	3
	266.1	0.5	10
Titanium	364.3	0.3	1
	365.4	0.15	2
	320.0	0.15	2
	375.3	0.15	3
	337.8	0.15	10
Vanadium	318.5	0.5	1
	306.0	0.15	5
	306.6	0.15	5
	439.0	0.5	10
	320.2	0.15	10

[a] The primary analytical wavelength has a relative value of 1; less sensitive lines have values greater than 1; more sensitive alternative wavelengths have values less than 1.

Table C.3 Preparation of 1 L of 1000 mg/L Standard Solutions for GFAAS (1–3,5–8)

Element	Procedure
Aluminum	Dissolve 1.000 g high-purity aluminum in concentrated HCl with heating; add a small drop of mercury (catalyst); filter to remove catalyst; dilute to volume.
Antimony	Dissolve 1.000 g antimony in nitric acid and dilute to volume. TOXIC
Arsenic	Dissolve 1.3203 g of arsenious oxide in 25 mL of 20% KOH. Neutralize with 20% H_2SO_4 and dilute to volume. TOXIC
Barium	Dissolve 1.437 g barium carbonate in 20 mL 1 + 1 nitric acid and dilute to volume.
Beryllium	Dissolve 1.000 g beryllium in 20 mL 1 + 1 nitric acid and dilute to volume. TOXIC
Bismuth	Dissolve 1.000 g bismuth in 1 + 1 nitric acid and dilute to volume.
Boron	Dissolve 57.20 g boric acid in deionized water and dilute to volume.
Cadmium	Dissolve 1.000 g cadmium in 50 mL 1 + 1 nitric acid and dilute to volume. TOXIC
Calcium	Dissolve 2.497 $CaCO_3$ in 50 mL water, dropwise add 10 mL nitric acid, and dilute to volume.
Chromium	Dissolve 1.000 g chromium in 50 mL 1 + 1 HCl and dilute to volume.
Cobalt	Dissolve 1.000 g cobalt in a minimum volume of 1 + 1 nitric acid and dilute to volume.
Copper	Dissolve 1.000 g copper in 50 mL 1 + 1 nitric acid and dilute to volume.
Gallium	Dissolve 1.000 g gallium in 20 mL aqua regia with heating and dilute to volume.
Germanium	Dissolve 1.000 g germanium with 5 mL HF in a Teflon beaker, add concentrated nitric acid to complete dissolution, and dilute to volume.
Gold	Dissolve 1.000 g gold in aqua regia, heat to dryness, dissolve residue in 50 mL HCl, and dilute to volume.
Indium	Dissolve 1.000 g indium in a minimum volume of 1 + 1 nitric acid and dilute to volume.
Iron	Dissolve 1.000 g high-purity iron in 50 mL 1 + 1 nitric acid and dilute to volume.
Lead	Dissolve 1.000 g lead (or 1.598 g lead nitrate) in 50 mL 1 + 1 nitric acid and dilute to volume.
Lithium	Dissolve 5.234 g lithium carbonate in 20 mL of 1 + 1 nitric acid and dilute to volume.
Magnesium	Dissolve 1.000 g magnesium in 20 mL 1 + 1 nitric acid and dilute to volume.
Manganese	Dissolve 1.000 g of manganese in 50 mL 1 + 1 nitric acid and dilute to volume.
Molybdenum	Dissolve 1.000 g molybdenum in 50 mL hot nitric acid and dilute to volume.

(continued)

Table C.3 (*continued*)

Element	Procedure
Nickel	Dissolve 1.000 g nickel in 50 mL 1 + 1 nitric acid and dilute to volume.
Phosphorus	Dissolve 4.3938 g KH_2PO_4 in deionized distilled water and dilute to volume.
Platinum	Dissolve 1.000 g platinum in 20 mL aqua regia, evaporate to dryness, redissolve in 5 mL of hydrochloric acid, and dilute to volume.
Selenium	Dissolve 1.000 g selenium in 20 mL 1 + 1 nitric acid and dilute to volume. TOXIC
Silver	Dissolve 1.000 g silver in 20 mL 1 + 1 nitric acid and dilute to volume.
Tellurium	Dissolve 1.000 g tellurium in 50 mL 1 + 1 nitric acid and dilute to volume. TOXIC
Thallium	Dissolve 1.303 g of thallium nitrate in 20 mL nitric acid and dilute to volume. TOXIC
Vanadium	Dissolve 1.000 g vanadium in 50 mL nitric acid and dilute to volume.
Zinc	Dissolve 1.000 g zinc in a minimum volume of nitric acid and dilute to volume.

Table C.4 Preparation of 100 mL of 10,000 mg/L (1%) Chemical Modifier Solutions for GFAAS (1–3,5–8)

Chemical Modifier	Procedure
Calcium	Dissolve 4.0942 g calcium nitrate in deionized water and dilute to volume.
Magnesium	Dissolve 10.5495 g $Mg(NO_3)_2 6H_2O$ in deionized water and dilute to volume.
Nickel	Dissolve 4.9545 g $Ni(NO_3)_2 6H_2O$ in deionized water and dilute to volume.
Palladium	Dissolve 1.000 g palladium in a minimum volume of aqua regia and dilute to volume; or dissolve 2.16 g palladium nitrate in 100 mL deionized water.
Phosphate	Dissolve 1.3905 g diammonium hydrogen phosphate in deionized water and dilute to volume.
Sulfuric acid	1 mL concentrated sulfuric acid is diluted to volume.

REFERENCES

1. *Analytical Methods for Atomic Absorption Spectrometry*, Perkin Elmer Corporation, Norwalk, CT, 1982.
2. *Recommended Conditions for THGA Furnaces*, Perkin-Elmer Corporation, Norwalk, CT, 1991.

3. *Atomic Absorption Volume II. Methods Manual for Furnace Operation*, Thermo Jarrell Ash, Franklin, 1993.
4. B. Radziuk, G. Rödel, M. Zeiher, S. Mizuno, and K. Yamamoto, *J. Anal. Atom. Spectrom.*, **10**, 415 (1995).
5. J. W. Robinson, *CRC Practical Handbook of Spectroscopy*, CRC Press, Boca Raton, FL, 1991.
6. W. Slavin, *Graphite Furnace AAS: A Source Book*, Perkin-Elmer Corporation, Norwalk, CT, 1984.
7. *Analytical Methods Manual for Flame Operation. Volume I. Operators Manual*, Thermo Jarrell Ash, Franklin, MA, 1986.
8. A. Varma, *CRC Handbook of Furnace Atomic Absorption Spectrometry*, CRC Press, Boca Raton, FL, 1990.

APPENDIX D

GLOSSARY

***a*:** *see* Absorptivity.

***A*:** *see* Absorbance.

A_{peak}: peak absorbance of a GFAAS measurement (unitless).

Absolute analysis: quantitative analysis from a theoretical equation and a single measurement of a sample.

Absorbance (*A*): negative logarithm of the transmittance; is directly proportional to the amount of analyte introduced into the furnace (unitless).

Absorptivity (*a*): constant that describes the sensitivity of an atomic line for AAS (L/g cm).

Accuracy: closeness between a measured value and the true or accepted value.

ac Zeeman: use of an electromagnet to achieve the Zeeman effect for background correction.

Allowed transitions: spectral transitions that have a high probability of occurring and therefore produce intense atomic lines.

Anomalous Zeeman effect: splitting of an atomic absorption line in a transverse magnetic field into more than two σ components and one π component.

Ash step: *see* Pyrolysis step.

Atom cell: device that converts a sample into gaseous atoms.

Atomic absorption (AA): absorption of light by an atom to promote an electron from a lower electronic level (usually a ground energy state) to a higher electronic level.

Atomic absorption cookbook: methods manual that accompanies commercial atomic absorption instruments.

Atomic absorption spectrometry (AAS): use of the atomic absorption phenomenon to quantify the concentration of elements in samples.

Atomic emission: following nonradiative (thermal) excitation of an atom, the deactivation of that atom from a higher electronic level to a lower electronic level by release of a photon.

Atomic fluorescence: following radiative excitation of an atom (absorption), the deactivation of that atom from a higher electronic level to a lower electronic level by the release of a photon.

Atomic spectroscopy: interaction of light with gaseous atoms.

Atomization step: portion of the furnace cycle in which the analyte is converted to atoms and the absorption measurement is performed. The temperature employed depends on the volatility of the analyte.

Atomization temperature optimization: series of absorbance measurements for a fixed concentration of analyte as a function of atomization temperature. Normally the lowest atomization temperature that gives complete atomization of analyte is considered optimum.

Aufbau order: relative energies of atomic subshells.

***b*:** *see* Pathlength.

Background correction: method to correct for background attenuation of source radiation. Common methods include nearby line, continuum source, self-reversal, and Zeeman background correction.

Baseline offset compensation: collection of data points just prior to atomization to compensate for drift between atomization cycles.

Boltzmann distribution: a mathematical relationship that gives the ratio of atoms in an excited state to the ground state.

$C_{A,Org}$: concentration of analyte in an organic phase.

$C_{A,W}$: concentration of analyte in an aqueous phase.

C_P: analyte concentration at the peak maximum in a flow injection analysis.

C_0: analyte concentration injected in a flow injection analysis.

Calibration graph (or curve): graph of integrated absorbance versus the amount of analyte in standard solutions. Generally the linear portion of the graph is used for quantitative analysis, and least squares analysis is used to determine a mathematical relationship between absorbance and concentration. The mathematical relationship is used to quantify the concentration of analyte in samples.

Calibration sensitivity (s): slope of a calibration graph.

Capacitively coupled plasma (CCP): argon plasma formed in a conventional atomic absorption graphite furnace by a graphite antenna coupled to a radiofrequency generator.

CCD: *see* Charge-coupled device.

CCP: *see* Capacitively coupled plasma.

Characteristic mass (m_0): mass of analyte that gives an absorbance value of $0.0044s$: $m_0 = 0.0044/s$, where s is the slope of the calibration graph. The characteristic mass is used to evaluate the performance of GFAAS instruments.

Charge-coupled device (CCD): multichannel solid-state detector, consisting of a two-dimensional array of solid-state devices on a silicon chip, which provides comparable sensitivity to a photomultiplier tube.

Char step: *see* Pyrolysis step.

Chemical interferences: reduction of the analyte absorbance signal by interaction(s) with one or more concomitant(s) compared to standards without the concomitant(s).

Chemical modifiers: reagents added to solutions for GFAAS analysis to reduce chemical interferences. A chemical modifier must be available in high purity, cannot degrade the lifetime of graphite tubes or platforms, and cannot induce spectral interferences. For volatile elements, the chemical modifier often serves to allow the use of a relatively high pyrolysis temperature to remove matrix components.

Clean step: portion of the furnace cycle in which the furnace is heated to a sufficiently high temperature to remove residual analyte. Typically temperatures of 2000–2700°C are employed.

Coefficient of variation: *see* Relative standard deviation.

Collisional broadening: width of an atomic line due to collisions with other gaseous species.

Collisional cross section (σ): area occupied by colliding particles.

Combustion: method of sample preparation in which the sample is heated to a sufficiently high temperature to vaporize its organic constituents, followed by extraction of the analyte with a mineral acid.

Continuum source atomic absorption: atomic absorption with a continuum source (usually a xenon arc lamp) as the light source. A high-resolution monochromator is required to obtain a sufficiently narrow bandpass to obtain comparable sensitivity to a line source.

Continuum source background correction: use of a continuum source (usually a deuterium arc) to measure and correct for spectral backgrounds.

Cool-down step: portion of the furnace cycle after the pyrolysis step in which the furnace temperature is decreased to ambient to allow more isothermal atomization. The furnace is then heated to the atomization temperature with maximum power heating. The use of a cool-down step creates a larger temperature gradient between the wall and the L'vov platform, resulting in more isothermal atomization and reduced chemical interferences.

CRA: carbon rod atomizer, a trademark of Varian Instruments.

***D*:** distribution ratio.

***D*:** dispersion.

D_c: distribution coefficient.

dc Zeeman: use of a permanent magnet with a rotating polarizer to achieve the Zeeman effect for background correction.

Decomposition bomb: system for wet decomposition of samples. The sample and reagents are placed in a bomb and heated in a muffle furnace.

Delayed atomization cuvette: graphite tube in which the center of the tube, onto which sample is deposited, is thicker than the ends of the tube, so that the temperature of the tube center is lower during the initial portion of the atomization step. This allows atomization of the sample to occur after the ends of the tube have reached the final atomization temperature, reducing chemical interferences.

Detection limit: *see* Limit of detection.

Deuterium arc background correction: *see* Continuum source background correction.

Direct solid sampling: direct insertion of solid materials into a graphite tube for analysis.

Direct Zeeman effect: application of a magnetic field to the light source for background correction.

Dispersion (*D*): degree of mixing of a sample with a flow stream in flow injection analysis.

Distribution coefficient (D_c): factor used to evaluate the efficiency of an ion-exchange procedure.

Distribution ratio (*D*): factor used to evaluate the efficiency of an extraction procedure.

Dry ashing: *see* Combustion.

Dry step: portion of the furnace program in which solvent is removed from the analyte. The furnace is usually heated slowly to a temperature just high enough to remove the solvent without spattering (typically 100–150°C for aqueous solutions).

***e*:** charge of an electron (1.6021×10^{-19}°C).

Échelle monochromator: high-performance monochromator that typically uses an échelle grating and a second dispersing element to separate overlapping orders of the grating onto a second dimension.

EDL: *see* Electrodeless discharge lamp.

Einstein coefficient for absorption (B_{01}): probability coefficient used to evaluate the rate of absorption (cm^3/J s Hz).

Einstein coefficient for spontaneous emission (A_{10}): probability coefficient used to evaluate the rate of spontaneous emission (s^{-1}).

Einstein coefficient for stimulated emission (B_{10}): probability coefficient used to evaluate the rate of stimulated emission (cm^3/J s Hz).

Electrodeless discharge lamp (EDL): light source for atomic absorption fabricated from a sample of the analyte in a quartz sphere surrounded by the coil of a microwave or radiofrequency generator.

Electrostatic precipitation: method to determine metals in air in which a high-voltage supply is used to capture particles from the air sample.

Enhancement factor: increase in sensitivity of a flow injection (FI) preconcentration system in which different conditions are used in the FI method and the standard procedure.

External gas flow: flow of purge gas outside the graphite.

f_{xy}: oscillator strength for a transition between states *x* and *y*.

FAPES: *see* Graphite furnace capacitively coupled plasma atomic emission spectrometry.

Fast furnace program: relatively short furnace program (<1 min) employed to increase the sample throughput of GFAAS.

FI: flow injection.

Fine structure: presence of two or more closely spaced spectral atomic lines due to differences in J values.

Flame atomic absorption spectrometry: use of the atomic absorption phenomenon with a flame as the atom cell for quantitative elemental analysis.

Flow injection (FI): for GFAAS, a method of preconcentration/separation that consists of a low-pressure column and a flow stream. The flow injection unit is typically synchronized to operate in parallel with the furnace cycle of the GFAAS instrument.

Forbidden transitions: spectral transitions with a low probability of occurring and consequently induce weak atomic lines.

Furnace atomization plasma emission spectrometry (FAPES): *see* Graphite furnace capacitively coupled plasma atomic emission spectrometry.

Furnace program: sample introduction and series of heating steps (typically dry, pyrolysis, cool down, atomization, and clean) employed in a GFAAS measurement.

Fusion: method of sample preparation in which the sample is mixed with an inorganic salt, placed in a crucible, and heated to a sufficiently high temperature to form a molten salt. The melt is subsequently poured into a dilute acid solution to release the analyte into solution.

g_x: statistical weight of state x.

Gas-phase interferences: gas-phase chemical reactions between vaporized analyte, either as a compound or the metal, with a concomitant, that prevents formation of analyte atoms.

GFCCP-AES: *see* Graphite furnace capacitively coupled plasma atomic emission spectrometry.

Glow discharge: atom cell commonly employed for the analysis of conducting solids consisting of two electrodes at a reduced pressure with a noble gas atmosphere.

Good laboratory practices: series of protocols to ensure high-quality data are produced by a laboratory.

Graphite furnace (GF): atom cell that includes graphite (usually a tube) that is heated resistively.

Graphite furnace atomic absorption spectrometry: use of atomic absorption with a graphite furnace as the atom cell for quantitative elemental analysis.

Graphite furnace capacitively coupled plasma atomic emission spectrometry (GFCCP-AES): atomic emission technique using a conventional graphite furnace with a capacitively coupled plasma. The graphite tube is heated with a conventional power supply, and the plasma is used to increase the number of excited-state atoms.

HCL: *see* Hollow cathode lamp.

Heisenberg uncertainty principle: impossibility of simultaneously specifying with arbitrary precision the position and momentum of a particle.

HG: *see* Hydride generation.

HGA: heated graphite atomizer, a tradename of Perkin-Elmer, Inc.

HG-GFAAS: *see* Hydride generation graphite furnace atomic absorption spectrometry.

Hollow cathode lamp: commonly used light source for atomic absorption composed of two electrodes inside a glass envelope containing a few torr of a noble gas. A hollow cathode is fabricated of the analyte, or a compound containing the analyte. A low-pressure discharge causes ejection of excited-state atoms. Radiative deactivation of these atoms causes the lamp emission.

Hydride generation (HG): conversion of the analyte to a volatile hydride that is then transported to an atom cell for measurement by an atomic spectrometry method.

Hydride generation graphite furnace atomic absorption spectrometry (HG-GFAAS): volatile hydrides are generated and trapped in a preheated graphite tube for GFAAS analysis.

Hyperfine structure: splitting of an atomic line due to an interaction between a nucleus with a nonzero orbital angular momentum and electrons.

Impaction: method to determine metals in air in which the air sample is passed through a jet onto the surface of a graphite tube, causing air particles to remain on the tube wall.

Integrated absorbance: area under a transient GFAAS signal, units of seconds. This is the preferred quantity to quantify absorbance for GFAAS.

Internal gas flow: flow of purge gas inside the graphite tube. This flow is turned off during the atomization step to obtain maximum sensitivity.

Internal standardization, method of: comparison of the detector response of the analyte to a second element (the internal standard) that has been added to standards and samples or to standards only if present in the sample. Ideally, the internal standard should be as similar to the analyte as possible (e.g., in terms of volatility). A multielemental instrument is required to use this technique, which has limited its application for GFAAS.

Inverse Zeeman: application of a magnetic field to an atom cell for background correction.

Involatile compound formation: reaction of the analyte with a concomitant to form a compound that is incompletely vaporized at the maximum possible atomization temperature.

Ion association complex: ionic compound composed of the analyte and a counterion that has high solubility in nonpolar solvents.

Isotope shift: splitting of atomic lines due to the presence of isotopes.

***j*:** total angular momentum quantum number of an electron.

***J*:** resultant angular momentum of an atom.

***jj* coupling:** coupling scheme in which spin–orbit coupling is assumed to be large, as in heavy ($Z > 30$) atoms.

***k*:** Boltzmann's constant (1.3805 × 10–25 J/K).

***K*:** spectroscopic constant involved in absolute analysis (cm^2/atom).

K_{ex}: overall extraction coefficient.

k_{yx}: rate constant (s^{-1}) for collisional deactivation from state y to state x.

***l*:** orbital angular momentum quantum number.

***L*:** resultant orbital angular momentum quantum number.

Laser ablation: method of sample introduction in which a solid sample is vaporized by laser light and transported by a stream of gas into an atom cell.

LDR: *see* Linear dynamic range.

Level of linearity (LOL): highest mass of standard on the linear dynamic range of a calibration graph. For GFAAS, the LOL is generally two to three orders of magnitude above the limit of detection.

Level of quantitation (LOQ): lowest mass of standard on the linear dynamic range of a calibration graph that can be determined quantitatively with an acceptable level of precision. The LOQ is generally a factor of 5 to 10 above the limit of detection.

Limit of detection (LOD): lowest mass of analyte that can be distinguished from statistical fluctuations of a blank, typically defined as $LOD = 3\sigma_B/s$, where σ_B is the standard deviation on the blank and s is the calibration sensitivity.

Linear dynamic range (LDR): portion of a calibration graph on which a linear relationship exists between mass of analyte and absorbance: $LDR = \log(LOL/LOQ)$, where LOL is the level of linearity and LOQ is the level of quantitation.

Linearization model: one of several mathematical algorithms developed to extend the linear dynamic range of GFAAS calibration graphs.

Line sources: light sources that emit narrow lines of light produced by electronic deactivation of atoms. Common line sources for atomic absorption include hollow cathode lamps and electrodeless discharge lamps.

LOD: *see* Limit of detection.

LOL: *see* Level of linearity.

Longitudinal Zeeman: orientation of magnetic field parallel to source light. Only σ (no π) components are present.

Longitudinally heated graphite furnace: furnace heated by a longitudinal flow of current. Significant temperature gradients exist in this design, which may allow condensation of analyte in relatively cool regions.

LOQ: *see* Level of quantitation.

L'vov platform: small graphite shelf onto which sample is deposited.

The platform is heated primarily radiatively, and hence atomization occurs after the tube has reached a relatively high, constant temperature, which reduces chemical interferences.

m_e: mass of electron at rest (9.1094×10^{-31} kg).

m_j: total magnetic quantum number.

m_l: orbital magnetic quantum number.

m_s: spin magnetic quantum number.

M_A: atomic mass of analyte (g/mol).

M_J: resultant magnetic quantum number.

M_L: resultant orbital magnetic quantum number.

$\overline{M_P}$: average atomic or molecular mass of a perturbing species (g/mol).

M_S: total spin magnetic quantum number.

$M_{A,W,n}$: mass of analyte in an aqueous phase after *n* extractions.

$M_{A,W,0}$: initial mass of analyte in an aqueous phase.

MAS: *see* Molecular absorption spectrometry.

Masking agents: reagents that complex with a metal and prevent its extraction into an organic phase.

Massmann furnace: *see* Longitudinally heated graphite furnace.

Matrix: all components of a sample except for the analyte.

Matrix effects: general term for interferences caused by concomitants in a sample matrix that change the absorbance signal compared to an equal amount of analyte in an aqueous standard.

Matrix modifier: *see* Chemical modifier.

Metal speciation: quantitative determination of chemical forms of a metal.

Microwave digestion: use of microwave energy to digest a sample. Advantages include rapid sample preparation and easy automation.

m_0: *see* Characteristic mass.

Modern furnace technology: series of instrumental and methodological developments that have been shown to provide greatest accuracy in GFAAS analyses. These developments include the use of an autosampler, pyrolytically coated graphite, furnace systems with high heating rates, platform, transversely heated graphite furnaces, Zeeman or self-reversal background correction, and chemical modifiers.

Molecular absorption: absorption of light by small molecules produced by the sample matrix.

Molecular absorption spectrometry (MAS): technique used to determine nonmetals, primarily halides, using conventional GFAAS instrumentation. A nonmetal (the analyte) and a reagent metal added in excess are introduced into the furnace. The absorbance of a diatomic molecule composed the analyte and the reagent is used to quantify the nonmetal.

Monochromator: wavelength selector that allows spectral scanning. Conventional échellette monochromators consist of two slits, two collimators, and a dispersing element (grating).

n: principal quantum number.

n: order of release of analyte from a surface.

n_P: population density of perturbers.

N_0: number of atoms introduced into an atom cell.

N_x: population density of state x (cm^{-3}).

Nearby line background correction: use of a second atomic line near the analytical line, which is not absorbed by the atoms, that is used to measure and correct for the background signal.

Normal Zeeman effect: splitting of atomic absorption lines in a transverse magnetic field into two σ and one π components.

Optical pyrometer: *see* Optical temperature sensor.

Optical temperature sensor: serves to monitor and control the temperature of the graphite tube during the atomization step.

Orbitals: locations to which electrons are assigned on the basis of quantum mechanics.

Orbital angular momentum quantum number (l): determines the shape of an atomic orbital.

Orbital magnetic quantum number (m_l): determines the orientation of an orbital in space.

Oscillator strength (f_{xy}): for a transition between states x and y, the ratio of an experimental or theoretical measurement of the transition probability divided by the classical transition probability.

Oxygen ashing: addition of oxygen into a graphite tube during the pyrolysis step to more completely break down the matrix.

Pathlength: distance the source beam travels through the atom cell in an absorption measurement (cm).

Pauli exclusion principle: in a given atom, only one electron may have a given set of four quantum numbers.

Peak absorbance: maximum absorbance value obtained during an atomization cycle.

Peak area: *see* Integrated absorbance.

Peak height: *see* Peak absorbance.

Percent transmittance (% T): percentage of source light that passes through an atom cell (%).

$pH_{\frac{1}{2}}$ value: pH value at which 50% of a metal is extracted.

Photomultiplier tube (PMT): most commonly used detector for GFAAS, consisting of several electrodes inside a quartz envelope maintained in a vacuum. Light hits a photoemissive cathode, releasing several electrons that are attracted to a more positive electrode (dynode), releasing several electrons per incident electron. This process is repeated with a total of 6–12 dynodes, allowing a possible gain of 10^6–10^7.

Physical interferences: changes in absorbance due to differences in the viscosity and surface tension of standard and sample solutions.

Platform atomization: deposition of sample onto a L'vov platform and its subsequent atomization from the platform.

Precision: repeatability of an analytical measurement, usually characterized by the standard deviation and relative standard deviation.

Probe: graphite shelf inserted into or removed from a graphite tube by a stepper motor.

Probe atomization: the deposition of a sample onto a probe and its subsequent atomization from the probe.

Polycrystalline graphite: low-cost, porous, reactive material for graphite tubes. This material is unsuitable for the determination of elements that form involatile carbides.

Principal quantum number (*n*): determines the energy of an orbital.

Pyrolysis step: portion of the furnace cycle in which matrix species that are more volatile than the analyte are removed by heating the tube. The temperature employed depends on the volatility of the analyte.

Pyrolysis temperature optimization: series of absorbance measurements for a fixed concentration of analyte as a function of pyrolysis temperature. Normally the highest pyrolysis temperature is employed without loss of analyte.

Pyrolytically coated graphite: polycrystalline graphite that has been coated with a thin layer (50 μm) of pyrolytic graphite. The relatively dense layer of pyrolytic graphite decreases the porosity and reactivity of polycrystalline graphite. This material is suitable for the determination of most elements by GFAAS.

***R*:** transition probability (J cm^3).

$\mathcal{R}$: ideal gas constant (J/K mol).

Recovery check: addition of analyte as standard to a sample to evaluate the accuracy of an analysis.

Relative standard deviation (RSD): quantitative measurement of the relative precision: $\mathrm{RSD} = S_D/\bar{x} \times 100\%$, where $\bar{x}$ is the mean of the measurements and S_D is the standard deviation of the measurements.

Residence time: time that gaseous analyte atoms are present in a graphite tube.

Resonance transitions: atomic absorption transitions that originate from the ground state.

Resultant angular momentum quantum number (*J*): describes the interaction of orbital and spin angular momentum of an atom.

Resultant magnetic quantum number (M_J): describes the orientation of total angular momentum vector in an atom with two or more electrons.

Resultant orbital angular momentum quantum number (*L*); determines magnitude of orbital angular momentum.

Resultant spin magnetic quantum number (M_S): describes orientation of spin angular momentum vector of an atom with two or more valence electrons.

Resultant spin quantum number (*S*): determines magnitude of spin angular momentum in an atom with two or more electrons.

Russell–Saunders (*LS*) coupling: assumes spin–orbit coupling is weak, as in the case of light ($Z < 30$) atoms.

s: *see* Calibration sensitivity.

s_D: *see* Standard deviation.

S: resultant spin quantum number.

Sample introduction: method of introducing a sample into an instrument for analysis.

Sample matrix: *see* Matrix.

Sample preparation: conversion of a sample into a form suitable for analysis.

Selection rules: rules derived from quantum mechanics that indicate the probability that a transition will occur.

Self-reversal: dip induced in the center of a spectral profile at high concentrations of analyte.

Self-reversal background correction: use of self-reversed hollow cathode lamp emission to measure and correct for background signals.

Shewhart control chart: plot of mean measured concentration of control samples versus time, which is used to monitor long-term precision and accuracy of an analytical method.

Slurry sampling: method of sample preparation in which a powdered sample is suspended in a liquid diluent and injected into the graphite tube.

Smith–Hieftje background correction: *see* Self-reversal background correction.

Solid sampling: *see* Direct solid sampling.

Speciation: *see* Metal speciation.

Spectral interferences: presence of concomitants that affect the quantity of source light that reaches the detection system.

Spectral overlaps: presence of a nonanalyte element that can absorb source radiation. These interferences are relatively rare with line source excited AAS.

Spectroscopy: interaction of electromagnetic radiation with matter.

Spin magnetic quantum number (m_s): determines orientation of the spin angular momentum vector.

Sputtering: vaporization and excitation of atoms from a solid due to collisions by ions in a low-pressure, electric discharge (e.g., a hollow cathode lamp).

SRMs: *see* Standard reference materials.

Stabilized temperature platform furnace (STPF) technology: *see* Modern furnace technology.

Standard additions, method of: calibration technique in which various amounts of analyte standards are added to aliquots of the sample. This procedure is often effective at compensating for chemical interferences that may not be accounted for by preparation of an aqueous calibration graph.

Standard deviation (S_D): quantitative measurement of precision of an analytical method:

$$s_D = \sqrt{\sum \frac{(x_i - \bar{x})^2}{n - 1}}$$

where x_i is an individual measurement, $\bar{x}$ is the mean of the measurements, and n is the number of measurements.

Standard reference materials (SRMs): samples that have been previously analyzed by two or more independent methods and are used to verify the accuracy of an analytical method.

Standardless analysis: *see* Absolute analysis.

Statistical weight (g_x): number of degenerate states that make up an energy level x.

STPF technology; *see* Modern furnace technology.

Structured background: background signals that vary with wavelength across the bandpass of a monochromator.

Term symbol: concise way to express quantum state of an atom, given by $^{2S+1}L_J$.

THGA: transversely heated graphite atomizer, a tradename of Perkin-Elmer, Inc.

Total angular momentum quantum number (j): describes the interaction of orbital and spin angular momentum of an electron.

Total magnetic quantum number (m_j): orientation of total angular momentum vector.

Transition probability (R): quantity used to evaluate the probability that a particular transition will occur (J cm^3).

Transmittance (T): fraction of source light that passes through an atom cell (unitless).

Transverse Zeeman: orientation of a magnetic field perpendicular to the direction of source emission. Both σ and π components are present.

Transversely heated graphite furnace: furnace heated by a transverse flow of current. There are reduced temperature gradients in the design, which reduces chemical interferences.

Two-line background correction: *see* Nearby line background correction.

Two-step furnace: graphite furnace system consisting of a graphite cup furnace and a transversely heated graphite tube furnace, each controlled by separate power supplies. Sample is deposited in the graphite cup and vaporized into the already hot graphite tube.

v: volume of scatters (cm^3).

V_{Org}: volume of an organic phase in an extraction.

V_W: volume of an aqueous phase in an extraction.

Voigt profile: description of the profile of an atomic line affected by Doppler and collisional broadening.

Volatile compound formation: chemical reaction between a concomitant with the analyte to form a relatively volatile compound that may be lost during the pyrolysis step.

$W_{x \to y}$ radiative rate of excitation per unit volume from state x to state y (cm^{-3}s^{-1}).

Wall atomization: deposition of sample onto the graphite tube followed by a conventional furnace program. Atomization occurs into a relatively cool gas that may induce chemical interferences.

Wavelength modulation: method of background correction in which the source is tuned to the analytical wavelength and away from the analytical wavelength.

Wet decomposition: use of mineral acids and oxidizing agents to dissolve a sample.

Xenon arc lamp: light source commonly used for continuum source atomic absorption composed of xenon at high pressure (10–30 atm) and two electrodes inside a sapphire envelope.

***Z*:** atomic number.

Zeeman background correction: use of a magnetic field (either ac or dc) to induce the Zeeman effect to measure and correct for atomic absorption background signals.

Zeeman effect: presence of atoms in a magnetic field, which causes atomic energy levels to split into two or more nondegenerate energy levels.

α_M: fraction of metal concentration present as uncomplexed metal.

β_n: formation constant of a metal chelate.

$\Delta\lambda_c$: width of an atomic line due to collisional broadening (nm).

$\Delta\lambda_D$: width of an atomic line due to Doppler broadening (nm).

$\Delta\lambda_V$: width of an atomic line predicted by the Voigt profile (nm).

ε_0: permittivity of free space (8.8854×10^{-12} $C^2/N\,m^2$).

λ: wavelength of light (nm).

λ_m: wavelength of maximum absorption.

ν: frequency of light (Hz).

ρ: energy density ($J/cm^3\,Hz$) of electromagnetic radiation at the frequency of a transition.

σ: collisional cross section (cm^2).

σ_B: standard deviation of a series of blank measurements.

σ_s: coverage of graphite surface by analyte (cm^{-2}).

τ: scattering coefficient.

$\%T$: *see* Percent transmittance.

INDEX